ACPL ITEM
DISCARDED

j52
Burrows, William E., 1937-
Mission to deep space

DO NOT REMOVE
CARDS FROM POCKET

ALLEN COUNTY PUBLIC LIBRARY
FORT WAYNE, INDIANA 46802

You may return this book to any agency, branch,
or bookmobile of the Allen County Public Library.

DEMCO

MISSION TO

DEEP SPACE

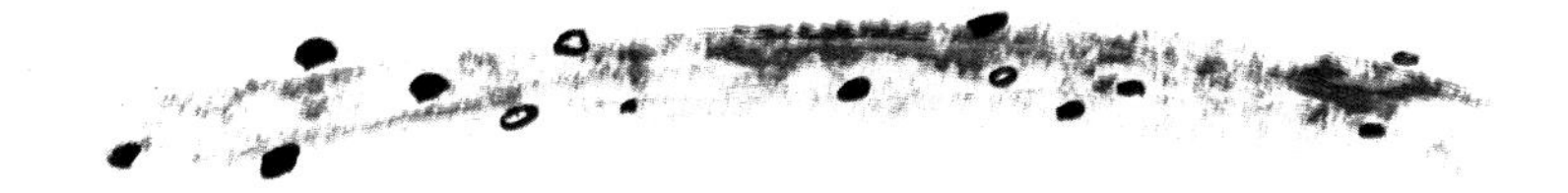

VOYAGERS' *Journey of Discovery*

MISSION TO DEEP SPACE

VOYAGERS' *Journey of Discovery*

by

William E. Burrows

Scientific American
BOOKS FOR YOUNG READERS

W. H. FREEMAN AND COMPANY • NEW YORK

Allen County Public Library
900 Webster Street
PO Box 2270
Fort Wayne, IN 46801-2270

Copyright © 1993 by William E. Burrows

All rights reserved.
No part of this book may be reproduced by any mechanical, photographic, or electronic process, or in the form of a phonographic recording, nor may it be stored in a retrieval system, transmitted, or otherwise copied for public or private use, without written permission from the publisher.

Printed in the U.S.A.

10 9 8 7 6 5 4 3 2 1

Library of Congress Cataloging-in-Publication Data

Burrows, William E., 1937-
MISSION TO DEEP SPACE : *Voyagers' journey of discovery* / by William E. Burrows.
p. cm.
Includes bibliographical references and index.

Summary: Describes the Voyager missions and what they revealed about the outer planets of our solar system.

ISBN 0-7167-6500-4

1. Outer space—Exploration—Juvenile literature. 2. Interplanetary voyages—Juvenile literature. 3. Voyager Project—Juvenile literature. [1. Voyager Project. 2. Planets—Exploration. 3. Outer space—Exploration.] I. Title.
QB500.22.B87 1993
523.4—dc20 92-29746
CIP

Contents

1

The Dream

Imagine traveling so far from Earth that the worlds you pass are made of fire and ice. You speed past a moon with cantaloupe skin and a blue streak running down its middle. Another moon looks like a pizza oozing molten sulfur instead of cheese. One moon has a huge "evil eye" that makes it look as sinister as a science fiction space station inhabited by dangerous life-forms. Still another is so shrouded by thick smog that it looks like a giant glowing orange.

Imagine streaking by one planet with terrible lightning and a violent storm that is hundreds of years old and larger than Earth itself; another that rotates sideways; and yet another whose rings look like a horseshoe and whose main moon is flying backward.

Well, all this really happened. It happened to a spacecraft named *Voyager* 2, which, with its sister craft *Voyager* 1, went on the greatest journey of exploration in history. *Voyager* 2 traveled to four giant planets: Jupiter, Saturn, Uranus, and Neptune; *Voyager* 1 flew by Jupiter and Saturn before heading out of our Solar System into deep space. The *Voyager* odyssey had a name as grand as the mission itself: the Grand Tour.

No human actually flew on *Voyager* 2 or *Voyager* 1. Yet in a sense, hundreds of scientists, engineers, technicians, and administrators in the National Aeronautics and Space Administration (NASA) and from the Jet Propulsion Laboratory (JPL) and other laboratories and universities around the world were explorers—at times even heroes. As a mission team they thought up and carried out the Grand Tour.

Some of the mission scientists were physicists, who study the physical laws of the universe: why things are the way they are. Others were meteorologists, who study the atmospheres of the planets and moons as meteorologists study Earth's atmosphere. They measured temperature, pressure, and wind direction. Others were geologists who wanted to learn about the surfaces of planets and moons and what happens beneath them. Still others were astronomers who were curious to know the exact positions of the planets and their satellites and how all of them interact.

These scientists had so many questions about the outer planets. . . .

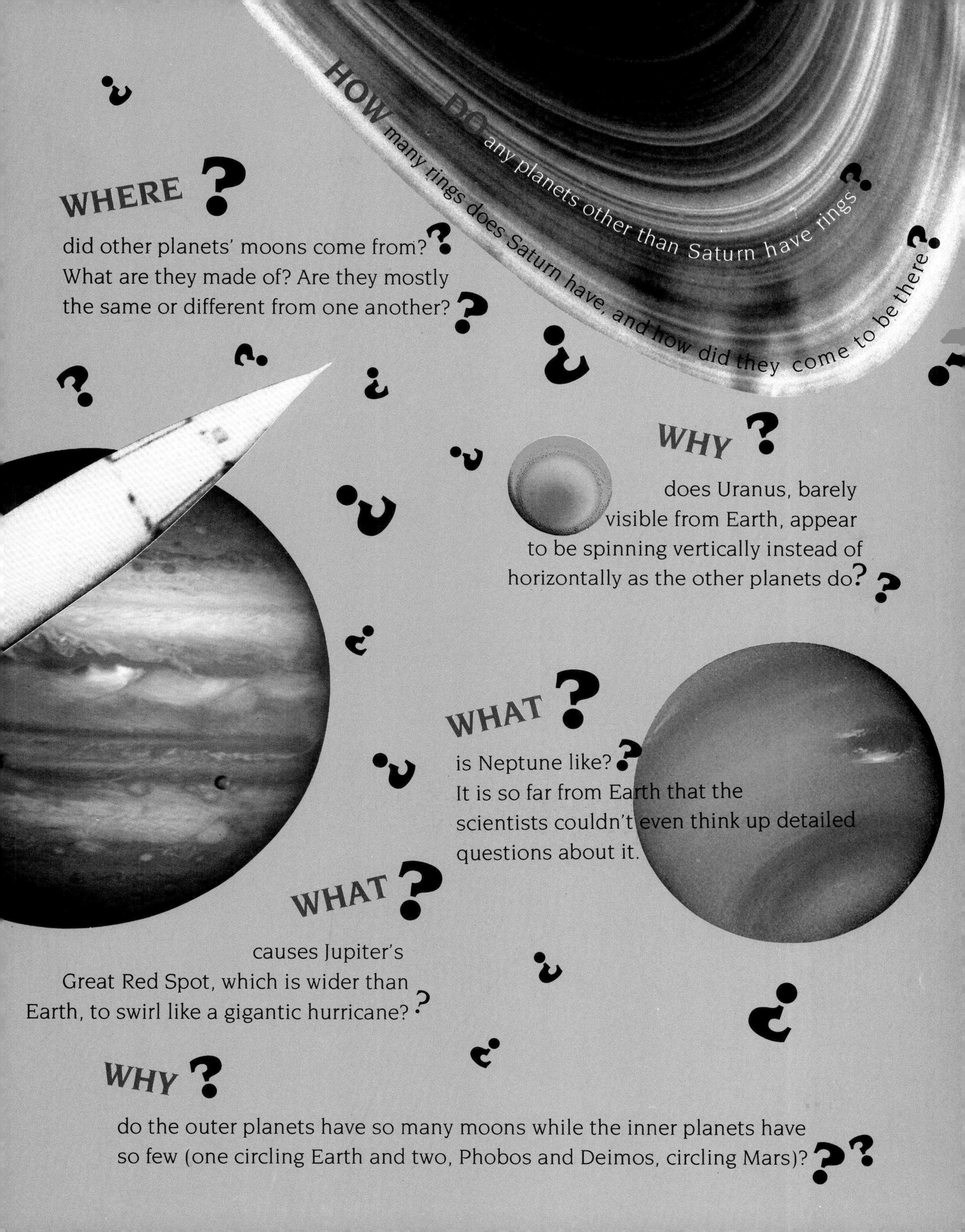

HOW many rings does Saturn have, and how did they come to be there?

DO any planets other than Saturn have rings?

WHERE ?

did other planets' moons come from? What are they made of? Are they mostly the same or different from one another?

WHY ?

does Uranus, barely visible from Earth, appear to be spinning vertically instead of horizontally as the other planets do?

WHAT ?

is Neptune like? It is so far from Earth that the scientists couldn't even think up detailed questions about it.

WHAT ?

causes Jupiter's Great Red Spot, which is wider than Earth, to swirl like a gigantic hurricane?

WHY ?

do the outer planets have so many moons while the inner planets have so few (one circling Earth and two, Phobos and Deimos, circling Mars)?

Planning the *Voyager* mission was a lot like planning a mission of exploration to the most remote areas of Earth. Both Antarctica and deep space are hostile environments. Just as antarctic explorers wear clothes that protect them from cold, the *Voyager* robot spacecraft needed protection from the interplanetary matter and cosmic radiation that they would encounter in space.

But exploring space is different from exploring Earth in a very important way. Unlike regions on Earth, which stay in place relative to one another, every destination in space is in constant motion. The simple two-dimensional maps that help Earth explorers get where they're going would not work in space. Trying to send a spacecraft to the Moon—or to Jupiter, Saturn, or Mars—is like trying to throw a football to a receiver who is running downfield. You have to throw the ball ahead of the player so that player and ball reach the same spot at the same instant. You have to know where the receiver will be when you want to pass and how much energy you will need to get the ball there. The *Voyager* team used computers and the changing, three-dimensional maps they create to figure out how to launch their craft.

Also, the project scientists had to figure out how to get spacecraft to travel the immense distances involved. At the beginning of the space age, it seemed obvious that the farther a spacecraft had to go, the more push it would need during launch. It is the first push, measured in pounds of thrust, that determines how fast and how far a spacecraft can travel. The heavier the spacecraft and the longer the distance, the more thrust the launch vehicle's rockets would have to produce. Pushing a spacecraft using fuel is called brute force propulsion. A robot explorer full of measuring instruments and carrying cameras would require an immense amount of fuel to send it where it was supposed to go.

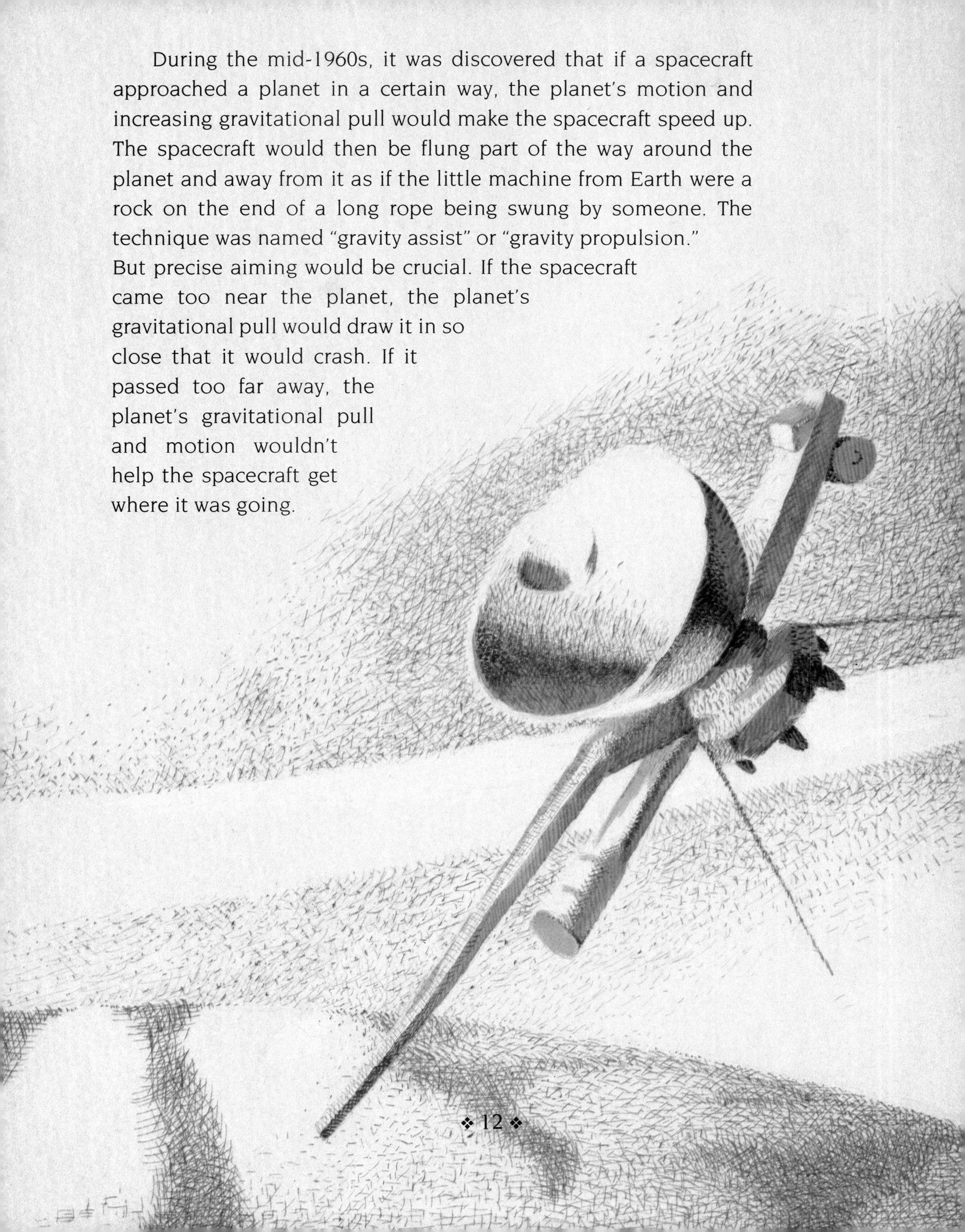

During the mid-1960s, it was discovered that if a spacecraft approached a planet in a certain way, the planet's motion and increasing gravitational pull would make the spacecraft speed up. The spacecraft would then be flung part of the way around the planet and away from it as if the little machine from Earth were a rock on the end of a long rope being swung by someone. The technique was named "gravity assist" or "gravity propulsion." But precise aiming would be crucial. If the spacecraft came too near the planet, the planet's gravitational pull would draw it in so close that it would crash. If it passed too far away, the planet's gravitational pull and motion wouldn't help the spacecraft get where it was going.

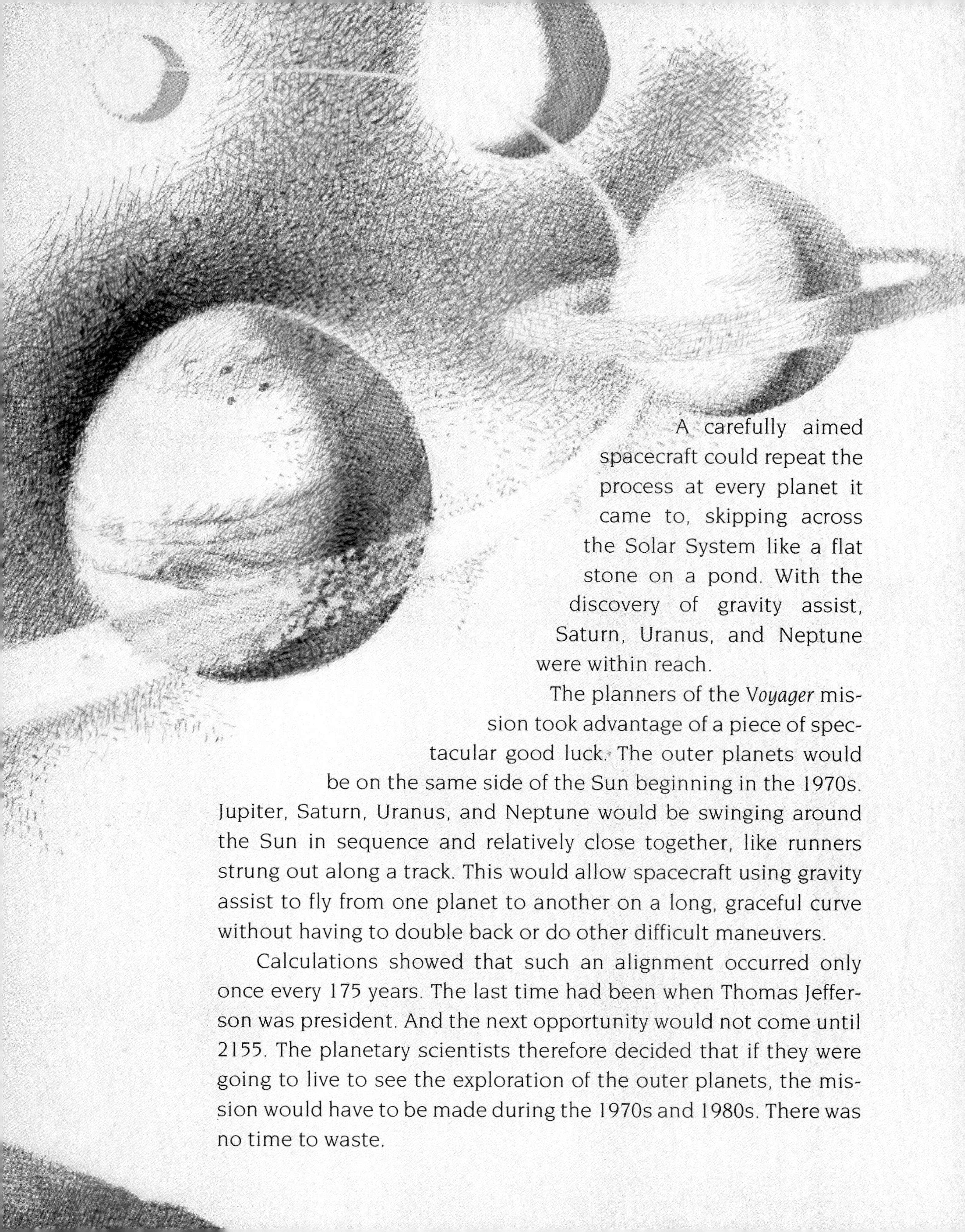

A carefully aimed spacecraft could repeat the process at every planet it came to, skipping across the Solar System like a flat stone on a pond. With the discovery of gravity assist, Saturn, Uranus, and Neptune were within reach.

The planners of the *Voyager* mission took advantage of a piece of spectacular good luck. The outer planets would be on the same side of the Sun beginning in the 1970s. Jupiter, Saturn, Uranus, and Neptune would be swinging around the Sun in sequence and relatively close together, like runners strung out along a track. This would allow spacecraft using gravity assist to fly from one planet to another on a long, graceful curve without having to double back or do other difficult maneuvers.

Calculations showed that such an alignment occurred only once every 175 years. The last time had been when Thomas Jefferson was president. And the next opportunity would not come until 2155. The planetary scientists therefore decided that if they were going to live to see the exploration of the outer planets, the mission would have to be made during the 1970s and 1980s. There was no time to waste.

2

The Spacecraft

Like other lunar and planetary explorers, the *Voyagers* did not have to fly through air, so they did not have to be streamlined like airplanes. Designing a spacecraft is like designing any other vehicle: what it looks like depends on what it's supposed to do. If you were asked to design a vehicle meant to carry 40 people and their luggage, you wouldn't design a sports car. If you have to move all that, the answer is to design a bus. The *Voyagers* had to function in a distant, hazardous environment. So they were designed to meet that challenge. They didn't have to look pretty, and they didn't. They looked like 827-kilogram (1,819-pound) insects.

LIFT OFF !
LIFT OFF !
LIFT OFF !
LIFT OFF !
LIFT OFF !

On August 20, 1977, *Voyager* 2 lifted off Launch Complex 41 at Cape Canaveral and headed for its rendezvous with Jupiter.

It was propelled by two powerful rockets, one fitted on top of the other. The bottom one was the Titan III-E launch vehicle, or booster. The top one, which would provide the final push, was the Centaur upper stage.

The Titan III-E's two solid boosters fired first. Their thunder shook the ground and seemed to pound the heads of the thousands of people who watched from a safe distance. A gigantic cotton ball of white smoke quickly enveloped most of the launch tower. The 15-story-high launch vehicle and its precious cargo rose past the top of the tower, gaining speed each second. Then it cleared the tower and arced out over the Atlantic Ocean on two pillars of fire.

At T plus 112 seconds (or 112 seconds after liftoff), the liquid-propelled engine in the Titan's first stage kicked in. Three powerful rockets were now pushing the Centaur far above the ocean, faster and faster. But the three mighty engines worked together for only ten seconds. Their propellant mostly used up, the two solid boosters disconnected from the core vehicle and fell away. Moments later they landed in the ocean and sank out of sight.

UP, UP, AND AWAY . . .

The first stage burned for 2 minutes and 26 seconds. Then, just before it too tumbled away, the Titan's second stage ignited. It burned for exactly 3 minutes and 30 seconds more. During that time, the Centaur's shroud opened like an immense mouth and fell away too.

Eight minutes after liftoff, the Titan III-E's second stage turned off and was also jettisoned. Now the Centaur's engine began its minute-and-a-half burn, putting it and its cargo in a "parking" orbit. During the next 30 minutes, the *Voyager* and the rocket that was going to give it its final push into outer space coasted in absolute silence halfway around Earth. Far below, tracking antennas and computers were busily gathering the final data that the mission scientists needed to send the robot explorer on a precise heading for Jupiter.

Fifty minutes after the Centaur left Florida, while it passed over the Indian Ocean, it came to life again. During the next six minutes it shoved *Voyager* 2 ever faster, until the spacecraft had gained enough velocity to break free of Earth's gravitational pull and begin its race to deep space. When the six minutes were up, Centaur separated in its turn, leaving *Voyager* 2's own little rocket motor to fire for 55 seconds more. One hour after liftoff, *Voyager* 2 was streaking at more than 35,000 km (22,000 miles) an hour, or 587 km (367 miles) a minute, toward the outer planets and then on to an unimaginable final destination.

(*Voyager* 1 was launched later but overtook its sister in the asteroid belt and arrived at Jupiter four months ahead of her.)

What Makes It Go?

Once the Voyagers were on their way, electricity had to keep the various systems powered. Those systems included computers, pumps, radios, tape recorders, navigation equipment, antennas, sensors (which looked and listened like eyes and ears), and other machines that would perform experiments. Power on a spacecraft can come from solar panels, which capture and convert the energy in sunlight into electricity, or from a nuclear generator, which converts the heat from plutonium into electricity. The Voyagers would travel too far from the Sun to use solar panels, so they had to run on nuclear energy.

The *Voyagers* had several tiny steering jets, called thrusters. These were fired in bursts as short as a fraction of a second in order to change the spacecraft's direction or simply to keep it stable. The thrusters fired a hypergolic fuel—one made of two chemicals that ignite by touching.

The *Voyagers* had to know where they were and where they were

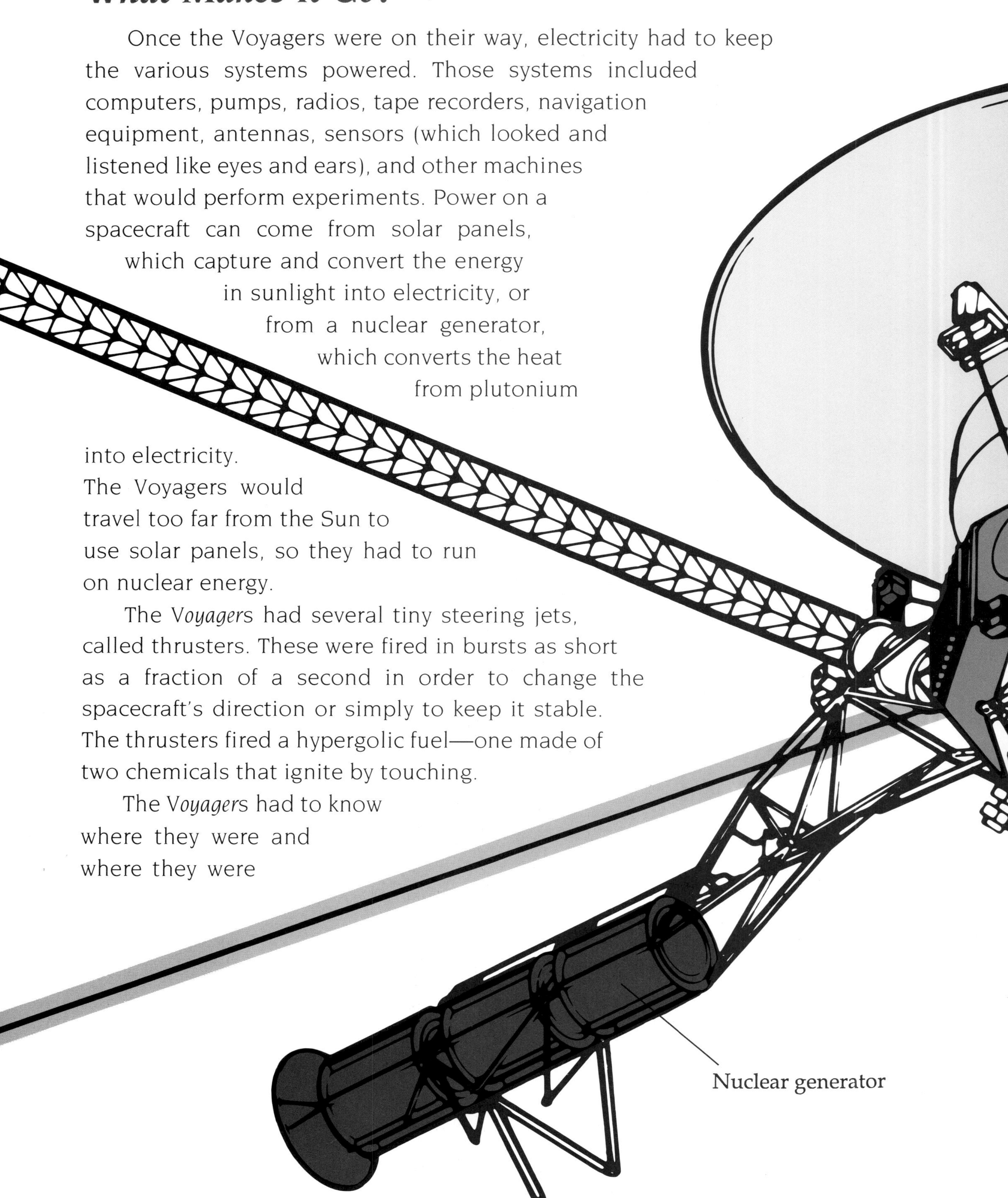

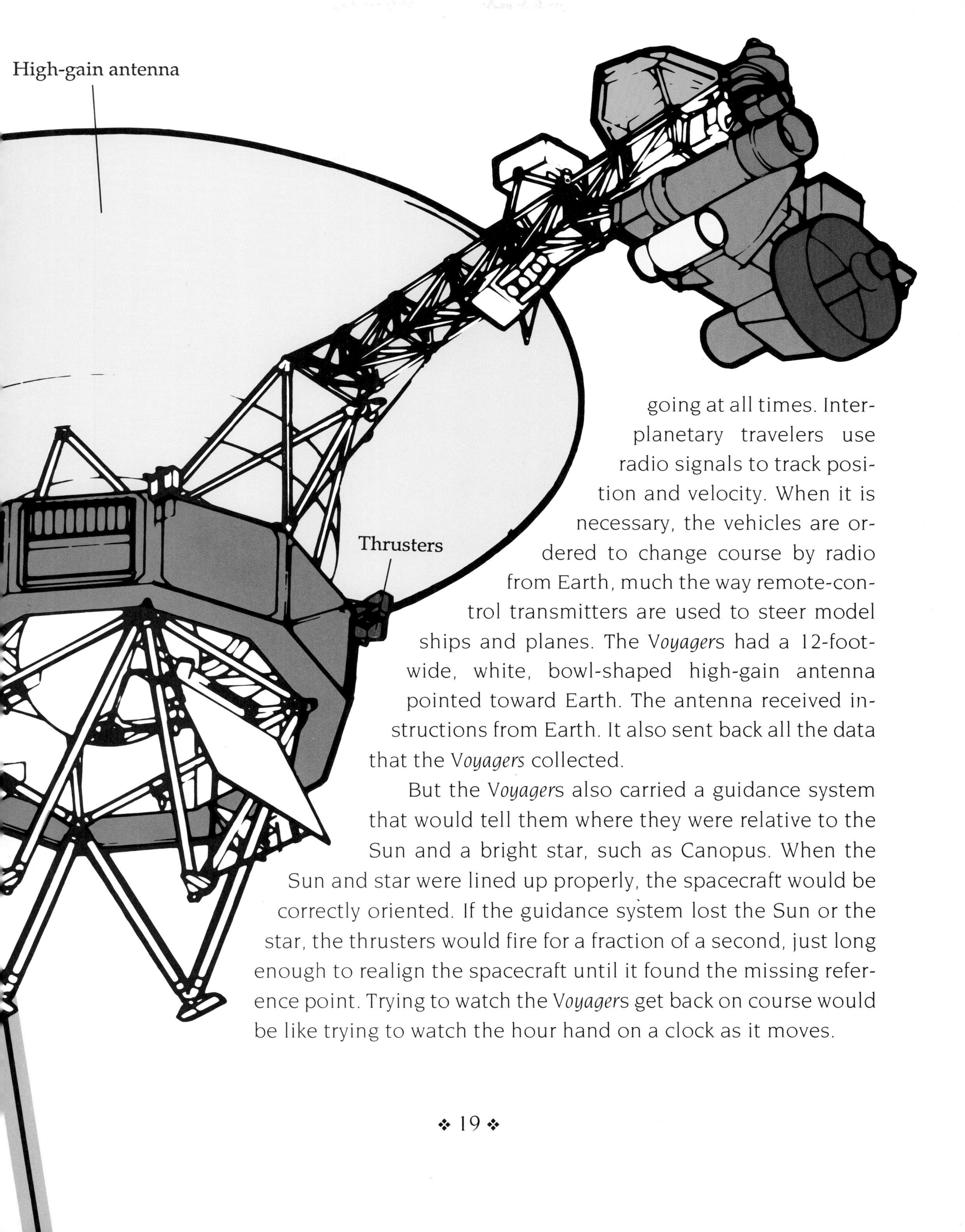

going at all times. Interplanetary travelers use radio signals to track position and velocity. When it is necessary, the vehicles are ordered to change course by radio from Earth, much the way remote-control transmitters are used to steer model ships and planes. The *Voyager*s had a 12-foot-wide, white, bowl-shaped high-gain antenna pointed toward Earth. The antenna received instructions from Earth. It also sent back all the data that the *Voyagers* collected.

But the *Voyager*s also carried a guidance system that would tell them where they were relative to the Sun and a bright star, such as Canopus. When the Sun and star were lined up properly, the spacecraft would be correctly oriented. If the guidance system lost the Sun or the star, the thrusters would fire for a fraction of a second, just long enough to realign the spacecraft until it found the missing reference point. Trying to watch the *Voyager*s get back on course would be like trying to watch the hour hand on a clock as it moves.

How Does It Do Its Work?

Each Voyager carried instruments for doing scientific investigations. Most of the instruments were mounted on a science boom, or arm, that stuck out of one end of the bus—the center of the spacecraft that holds most of the scientific equipment.

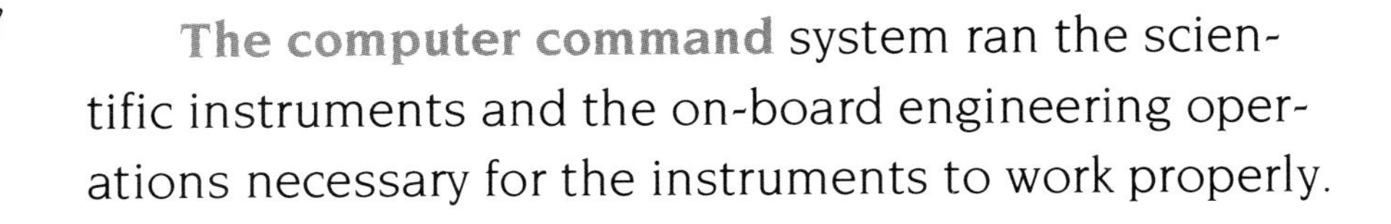

The computer command system ran the scientific instruments and the on-board engineering operations necessary for the instruments to work properly.

The steerable scan platform was like a head at the end of a long neck, moving up and down and from side to side to point the cameras in the right direction. On the scan platform were a wide-angle camera for photographing close up and a narrow-angle camera for photographing from far away.

The cameras broke down each picture into millions and millions of bits and pieces of information and sent them back to Earth, where they could be put back together to make a whole picture.

The radio antenna was a radio amplifier attached to a pair of 33-foot-long "rabbit ears" like the ones on top of old television sets. It listened for radio signals from the Sun, the planets, the planets' magnetic fields, and the kind of lightning found on Jupiter.

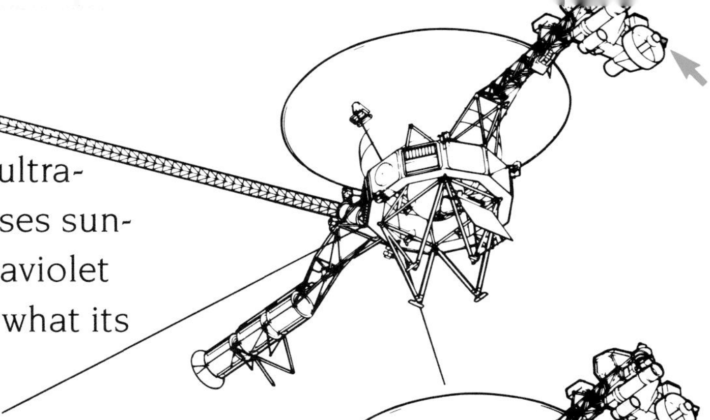

The ultraviolet spectrometer measured ultraviolet light. Ultraviolet light from the Sun causes suntans and sunburns. Measuring how much ultraviolet light comes from a planet helps us figure out what its atmosphere is made of.

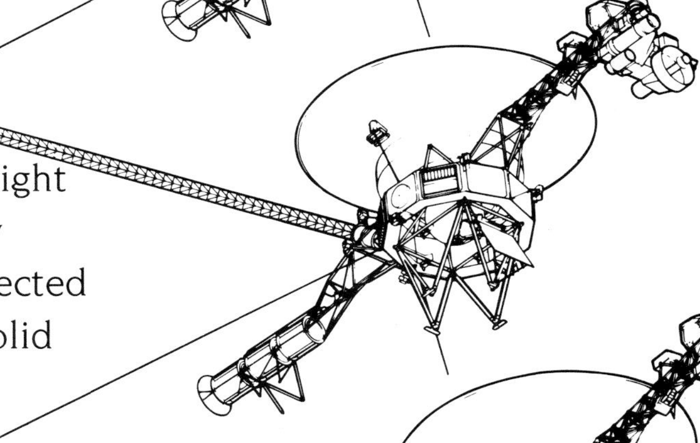

The light change detector measured how light changed as it was reflected from or absorbed by planets, moons, rings, and other objects. It collected information about the texture and makeup of solid surfaces and rings.

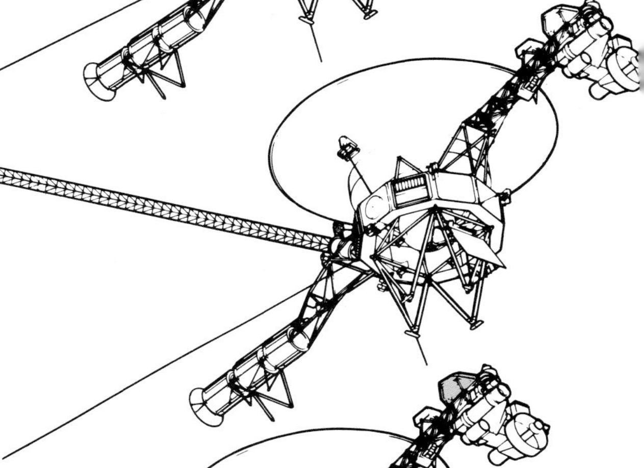

The magnetometers were used to detect and measure each planet's magnetic field, how the moons and rings reacted with the magnetic fields, and how the Sun's own magnetic field changes over distance. They were mounted far out on the booms so they wouldn't get interference from the magnetic field of the spacecraft.

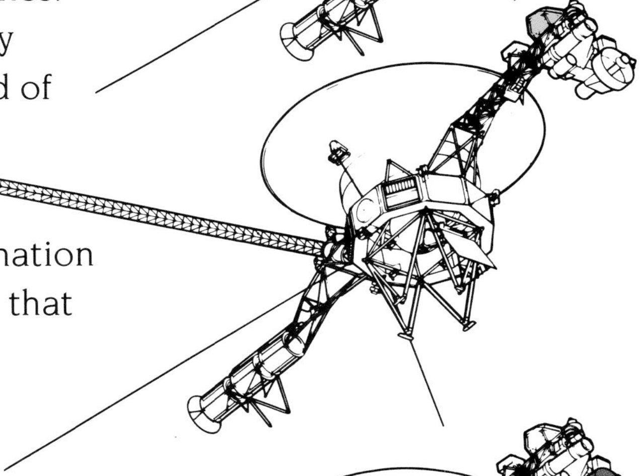

The plasma wave detector collected information about plasma—a soup of electrons and protons that is in the space between the planets.

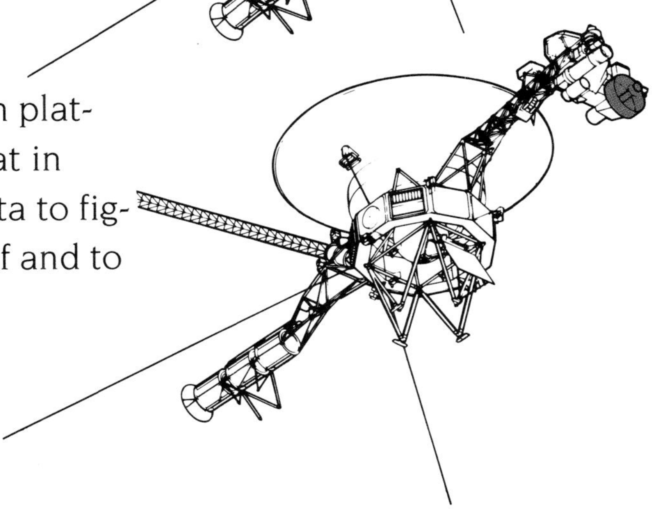

The infrared camera was also on the scan platform. This instrument can see variations in heat in fine detail. Mission scientists used infrared data to figure out what Jupiter's atmosphere was made of and to map the temperatures of its moons.

FOCUS ON:

A *Greeting to the* Universe

Each *Voyager* spacecraft carried a detailed message, describing many things on Earth, to any intelligent being who might find it—a greeting from the people of Earth to the entire universe. The message was recorded on two gold-plated phonograph records and attached to one side of each spacecraft, along with a cartridge and needle and a diagram of playing instructions. On the record were samples of 56 languages spoken on Earth and Earth sounds such as animal calls, laughter, a tractor, surf, rain, a jet plane, Morse code, and even a kiss. There were recordings of classical music, folk music from around the world, and popular songs.

There was also video coding for pictures of dozens of Earth things: scenes of humans working and playing, natural objects, and tools of art and science that define Earth as a special place. There were images of a seashell, a snowflake, a leaf, a supermarket, the Great Wall of China, an airplane in flight, the city of Boston, an X ray of a hand, Bushmen hunters, dolphins, a fetus, cells and cell division, the inside of a factory, the inside of a modern house, the United Nations building by day and night, people eating, licking, and drinking, Olympic runners, a Japanese schoolroom, children looking at a globe of Earth, scientists in a forest with chimpanzees, and an astronaut in space. The idea was to show an intelligent being in some far-off world something about what Earth looked like and what was happening to it in the second half of the twentieth century (though nothing bad such as war, crime, or pollution was shown). There was even a handwritten greeting in scribbly English from an Earth child to another inhabitant of the universe.

Hello from the children of the planet earth.

On December 10, 1978, when *Voyager* 1 was still almost 83 million kilometers (52 million miles) away from Jupiter, it began taking pictures of the giant planet that were better than any ever taken from Earth. In the days that followed, scientists studied the images with growing excitement.

For the scientists back on Earth, watching Jupiter day after day was like being in the spaceship itself. They could see the planet expand until it filled their computer screens. Since they could now see places they had never seen before, the scientists could pick out those of particular interest—the Great Red Spot, the belts of color, and the largest moons—for closer looks.

Now the two spacecraft—a pair of mechanical insects speeding toward a mountain-size world—picked up velocity as Jupiter's gravitational pull took increasing hold on them. On the day of closest approach, *Voyager* 1 streaked past Jupiter and its moons at 100,000 kilometers (62,000 miles) an hour, or about 1,500 kilometers (1,000 miles) a minute. (At that speed it could have flown around the Earth in less than half an hour.)

3
The King of the Sky

What We KNEW

A great deal was already known about Jupiter before the *Voyagers* arrived.

Jupiter's dimensions and mass had been calculated with great accuracy. Jupiter's mass is more than twice the total of all the other known planets in the solar system! If Earth is the size of a marble, Jupiter is the size of a softball.

Before *Voyager* we knew that Jupiter is made mostly of hydrogen and helium with some ammonia and methane. We knew that it gives off two to three times as much heat as it absorbs from the sun. We knew that Jupiter has colored belts that are wind currents created by its fast rotation. Jupiter rotates so quickly that its poles are flattened out a little.

Far larger than Earth itself, the Great Red Spot looks like an immense storm. As the years passed, it changed color,

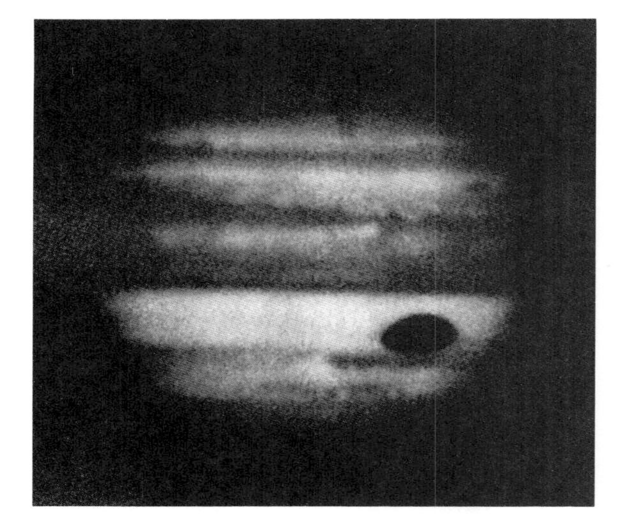

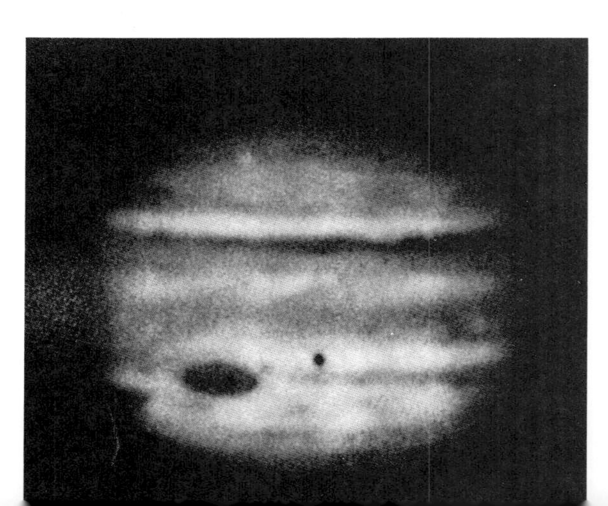

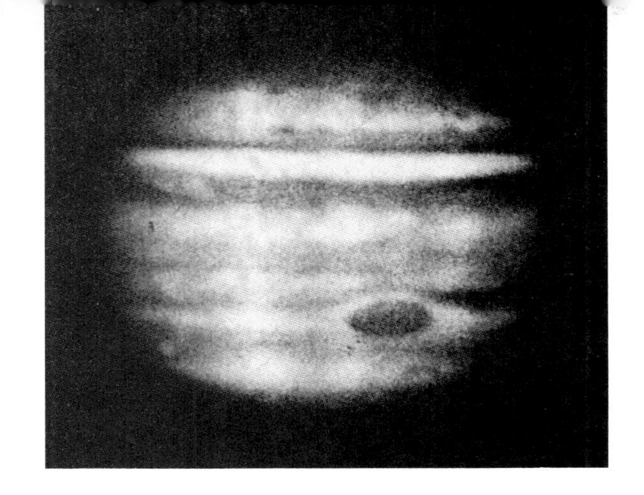

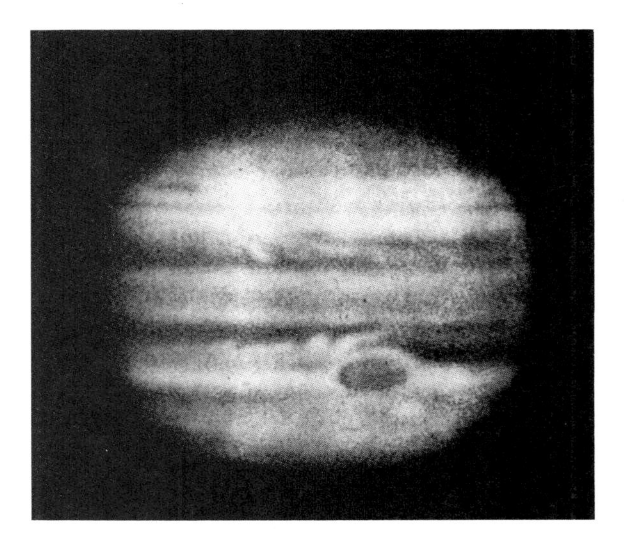

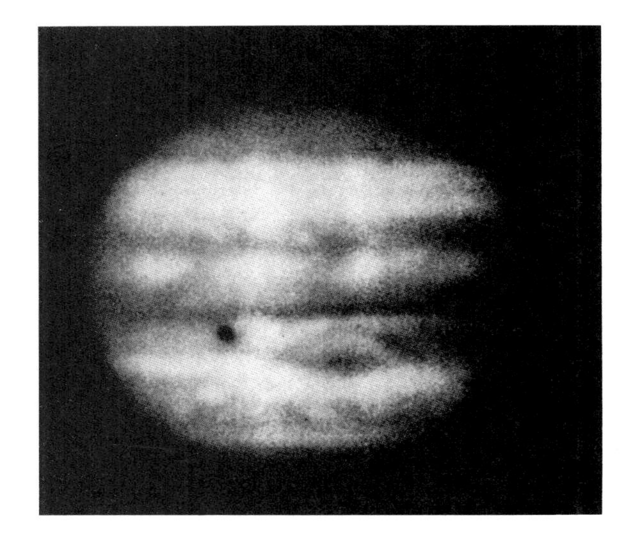

shape, intensity, and motion, just the way storms do on Earth. Once, it even seemed to disappear, only to return years later. But unlike storms on Earth, the Great Red Spot never seems to run out of energy.

When *Voyager* 1 arrived at Jupiter, 13 moons were already known. The four largest had been discovered by Galileo. We knew that two of those, Callisto and Ganymede, are solid and have polar caps.

The largest of the planets was well named. In Roman mythology Jupiter was the king of the gods and therefore the ruler of the sky. People had watched the giant planet long before the invention of the telescope. It reflects the Sun's light with a brightness second only to Venus, and under the right conditions, it can even cast shadows like our Moon. Telescopes provided a great deal of detailed information about the colossus of the night. But seeing more of Jupiter raised even more questions.

QUESTIONS, QUESTIONS, QUESTIONS

Every known led to one or more unknowns, and from there to still more questions.

Why do the belts seem to flow in opposite directions and change color?

If Saturn has rings, why doesn't Jupiter, Saturn's sister planet, have them too?

If Jupiter has its own rings, as some astronomers suspected, why are they not visible from Earth?

The biggest question of all had to do with the planet as a whole and with its moons. Jupiter is made almost entirely of hydrogen and helium. So is the Sun. Jupiter also gives off more heat than it absorbs just the way the Sun does. Most or all of Jupiter's moons seemed to be solid; so are the four inner planets. Jupiter's four largest planets vary in size from as big as Earth's Moon (Europa) to nearly as large as Mars (Ganymede). They are worlds of their own.

For all of these reasons, Jupiter and its moons are a lot like a miniature solar system. It's hard to study our whole Solar System because it is so big. The *Voyager* team hoped that studying Jupiter and its moons would tell them how solar systems work. That would help scientists think about our whole Solar System.

Many questions did not even occur to the *Voyager* scientists until the spacecraft got close to the planet. Would you be able to make up questions about a place you had never seen clearly before you got there? You can ask why people in the far north build their houses out of ice and eat whale blubber if you have seen them do it. Otherwise, you won't even think of the questions. Explorers do not know what they will find when they reach their destination. That's why they explore.

What We LEARNED

As the *Voyagers* flew by Jupiter, millions of bits of data a day were streaming into the Jet Propulsion Laboratory's computers from the scientific instruments aboard the spacecraft. Like an endless line of railroad boxcars, they came in together, were sorted according to their scientific cargo and then switched to their own special tracks. Scientists on all of the teams then analyzed what the bits of data carried and reported the findings to all the other scientists on the *Voyager* project.

Analyzing this data, the scientists dicovered that there are three narrow, dusty rings around Jupiter. Jupiter's rings have a bright, narrow segment within a broader, dimmer segment.

A great deal of data was also sent back about Jupiter's complex atmosphere, including the Great Red Spot, which flows counterclockwise and interacts with nearby, smaller spots like square dancers doing the do-si-do.

In addition, *Voyager* 1 returned unprecedented data on all the Jovian moons. Europa was seen to have a remarkably smooth terrain with very few craters, while four basic types of terrain were seen in images of Ganymede. And while photographing the rings, *Voyager* 2's cameras picked up a tiny new moon orbiting just off their outer edge. It was later called Adrastea. Close examination of the *Voyager* imagery would reveal two more baby moons, Metis and Thebe, bringing the known total to 16.

The *Voyager* mission's greatest discovery during the encounter with Jupiter involved Io—one of the four moons first discovered by Galileo. The first close-up peek at Io was startling. Seen in false color, which highlighted many of its features, there were orange, white, and yellow landscapes that even the imaging team could only describe as "grotesque," "diseased," "gross," and "bizarre."

But there was more. Early on the morning of March 8, *Voyager* 1 looked back at Io from a distance of 4 million kilometers (two and a half million miles) and took a long-exposure picture of Io's limb—its round edge—against a black background sprinkled with stars. The idea was to use the moon and the stars to help keep the spacecraft on course. Yet when Linda Morabito, an optical navigation engineer, studied the picture later that day, she noticed a crescent-shaped cloud extending 150 kilometers (100 miles) over Io's horizon. Since Io was known to have no atmosphere, the cloud made no sense.

At about the same time, other scientists had been studying hot spots at several places on Io. A comparison of data soon showed that one of the infrared team's hot spots was situated just about where one of the imaging team's plumes was spurting. The clouds must have been caused by an enormous volcanic eruption. Before the week was over, a total of eight volcanoes were located on *Voyager* 1's pictures of Io. Eventually, a ninth would be spotted.

The volcanoes solved a puzzle. Scientists had wondered why Io had no impact craters from objects that crashed into it. The other Galilean moons did. The eruptions from the volcanoes would have filled the craters with molten sulfur or another sulfuric material, giving the moon what planetary geologists call a new surface. Millions of years of such volcanic flow coated Io's surface until it looked like a pizza splotched with tomato sauce and hot mozzarella cheese.

The volcanoes on Io are a wonderful example of discoveries that no one expected. The spacecraft weren't programmed to look for them. But there they were, a complete surprise!

FOCUS ON IO

No Anchovies Please!

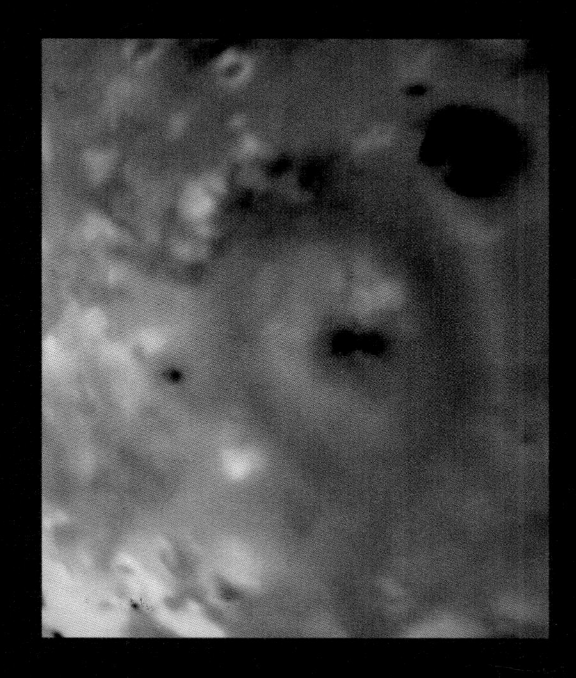

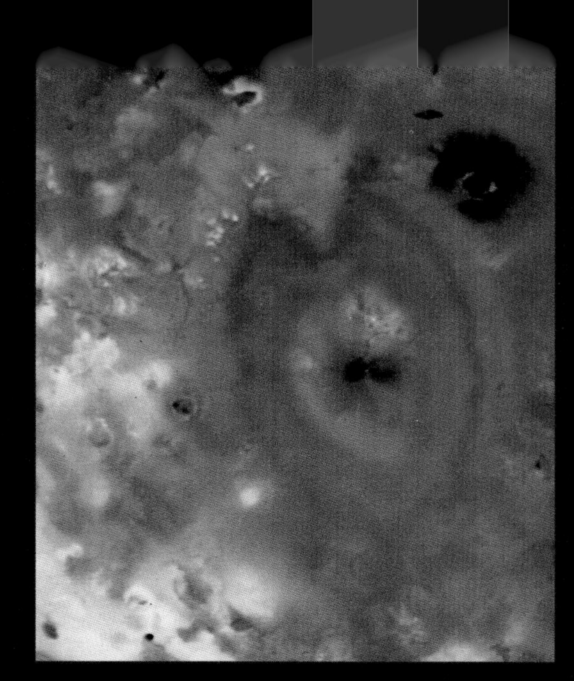

The volcanos on Io were erupting so violently that this one changed shape in the time between Voyager 1 and Voyager 2 flybys.

Prometheus, an active volcano on Io.

4

Sizing Up the Lord of the Rings

Both *Voyager* 1 and *Voyager* 2 used gravity assist to sling around Jupiter and gain speed as they headed toward Saturn.

On October 1, 1980—six weeks before nearest encounter—*Voyager* 1's cameras sent back pictures of Saturn and its rings with a resolution of about 990 kilometers (620 miles). This means the resolution was three times better than that of any pictures ever taken from Earth. The pictures were so sharp that Saturn's rings could be seen as complex bands that rippled out from the planet, and not as the flat sheets of material that scientists had thought they were. The rings are most likely made of the billions of fragments of a smashed moon.

What We KNEW

To the ancient Assyrians, Babylonians, Greeks, and others who had followed Saturn's course through the night sky, the majestic planet seemed to hang at or near the outer limit of the world. Galileo first trained his telescope on Saturn in July 1610 and was mystified by what appeared to be three distinct objects: a large one with a small one close by on either side. His drawing of what he saw looked like a round face with very large ears. He concluded that the smaller objects were moons, similar to the four he had spotted around Jupiter.

In 1655, using a better telescope, 26-year-old Christian Huygens saw that the objects were not moons at all. They were a continuous ring of bright material that surrounded the whole planet. And like the planet itself, the ring was tilted at 27 degrees. Huygens also discovered Titan, Saturn's largest satellite, which is bigger than the planet Mercury.

During the next century, Saturn, its growing number of known moons (nine at the start of the space age), and its ring system were studied through increasingly powerful telescopes. Saturn is the second-largest planet—95 Earths would fit inside

it—and it is made almost entirely of hydrogen and helium. But while Saturn is 95 times as large as Earth, its colossal gaseous body makes it proportionally much lighter than the home planet. This means that, like a gigantic wooden ball, Saturn is very large but not very dense. In fact, Saturn would float in water! Its low density and rapid rotation rate—a day on Saturn is only 10 hours, 14 minutes—make it bulge at the equator the way a spinning water balloon bulges in the middle.

But even the most powerful telescopes on Earth could not provide all the details scientists wanted. Saturn was simply too far away: 650 million kilometers (400 million miles) farther away from Earth than Jupiter is.

What We LEARNED

As *Voyager* 1 approached Saturn at more than 48,000 kilometers (30,000 miles) an hour, the pictures it was taking showed that a bonanza of new data lay ahead.

The Cassini Division, a gap in the middle of the rings, originally thought to be empty space, suddenly was seen to contain at least four of its own distinct bands. And a closer look would show that they, in turn, splintered into 20 or more even thinner bands. Scientists also speculated that the gaps in the Cassini Division were caused by very small "moonlets" that once plowed into the particles in their path, clearing them away as a snowplow clears snow.

What had once appeared to be a set of only a few major rings turned out on closer inspection to be more than a thousand smaller, threadlike rings, many of them made of tiny particles, all of them constantly, subtly changing. Some ring particles are as small as specks of dust, some as large as boulders. Mysterious dark lanes radiate across one of the rings like the spokes of a wheel. Small new moons act as shepherds to keep the rings together. The planet itself, which seemed from Earth to have a rather calm atmosphere, has several light and dark spots, each of which is thought to be a storm system thousands of kilometers wide. It has a dense core of rock, iron, and other heavy elements that are so hot they're liquid.

The outermost ring, the F ring, turned out to be three rings that formed a complex braided structure having a kink, or knot. The laws of motion and gravity suggested that whatever the thickness and composition of rings, they should certainly orbit a planet in a uniform pattern. The F ring seemed to defy the laws of nature. But now the scientists were not so sure they knew what the laws were. They might have to come up with new ones to explain the F ring.

Voyager 1's close-up imagery showed Saturn's atmosphere and cloud structure in great detail. The clouds were seen to be made mostly of ammonia ice crystals. Scientists got unprecedented views of the planet's swirling belts, which, unlike Jupiter's, all move in the same direction with respect to the equator. They also saw great fierce storms in considerable detail.

SATELLITES

Titan

Titan is Saturn's giant cloud-shrouded moon, the second largest in the Solar System after Jupiter's Ganymede. It is especially interesting to the scientists because it has the thickest atmosphere of any moon in the Solar System—so thick, in fact, that some investigators suspect that Titan is very much like Earth was several billion years ago.

Voyager scientists determined that Titan not only has a solid surface under its orange smog, but possibly oceans of condensed ethane as well, some more than a half mile deep. A human being floating on such a sea would be in a cold, dark place, where the temperature did not change much and where the distant Sun would penetrate the atmosphere just enough to cause an eerie orange glow. The areas of solid land are probably pockmarked with impact craters.

The winner of the Scariest-Looking-Moon-in-Our-Solar-System Contest, Mimas, innermost of Saturn's moderately large moons, was hit by an object that left a huge crater measuring 80 miles across. The crater dominates Mimas and seems to stare out at the rest of the universe.

The boundary between Tethys's heavily cratered regions and the more lightly cratered regions suggests early internal activity—perhaps volcanoes—that resurfaced the older area.

Hyperion is Saturn's strangely shaped, weird moon.

Enceladus is one of the most mysterious and frustrating bodies in the Solar System. The surface is relatively free of craters, which suggests the recent presence of water.

Mimas
Hyperion
ethys
Titan
Enceladus

FOCUS ON:

Long-Distance Repair. . . .

In the early morning of August 26, back at the Jet Propulsion Laboratory, elation suddenly turned to gloom. The *Voyager* 2's scan platform was frozen in its side-to-side track. This meant its two cameras, two spectrometers, and photopolarimeter could move up and down but not sideways. *Voyager* 2 sent a stream of blank imagery back to JPL. And the platform had stopped in a position where sunlight could damage the sensors themselves.

There was no time to panic. The control team sent *Voyager* 2 a message that put the instruments into standby mode. *Voyager* 2 sent back a signal telling its controllers that it had obeyed them.

Scientists didn't know what caused the scan platform failure. But they decided to try to gradually work the platform back and forth, the way you move a stiff arm to loosen its muscles. Little by little, the mechanism began to respond to commands from Earth. Finally, as *Voyager* 2 left the Saturnian system, it took fine wide-angle shots of the beautiful planet.

After studying the problem, members of the *Voyager* team concluded that the platform probably got stuck because it lost lubricant by scanning too quickly. If the platform was restricted to slow scans, the engineers concluded, there was no reason why *Voyager* 2 could not continue to Uranus and Neptune and collect excellent pictures.

So as the next leg of *Voyager* 2's odyssey—the four-and-a-half-year flight to Uranus—began, its managers, engineers, and scientists shared high expectations and nagging fears. Their spacecraft was headed where no other had gone before. *Voyager* 2's eyes could provide what no human had ever been able to see. But the little messenger from Earth would be eight and a half years into its mission and almost three and a half billion kilometers (two billion miles) from home when it reached Uranus. That would make it a decidedly old spacecraft by the standards of the space age. Worse, it would arrive almost totally deaf, partly blind, crippled, somewhat feeble, and perhaps even battered by debris through which it was flying at 50,000 kilometers (30,000 miles) per hour.

The mission scientists and engineers went to work. They had a lot to do. From billions of kilometers away, they needed to get *Voyager* 2 ready for the encounter with Uranus.

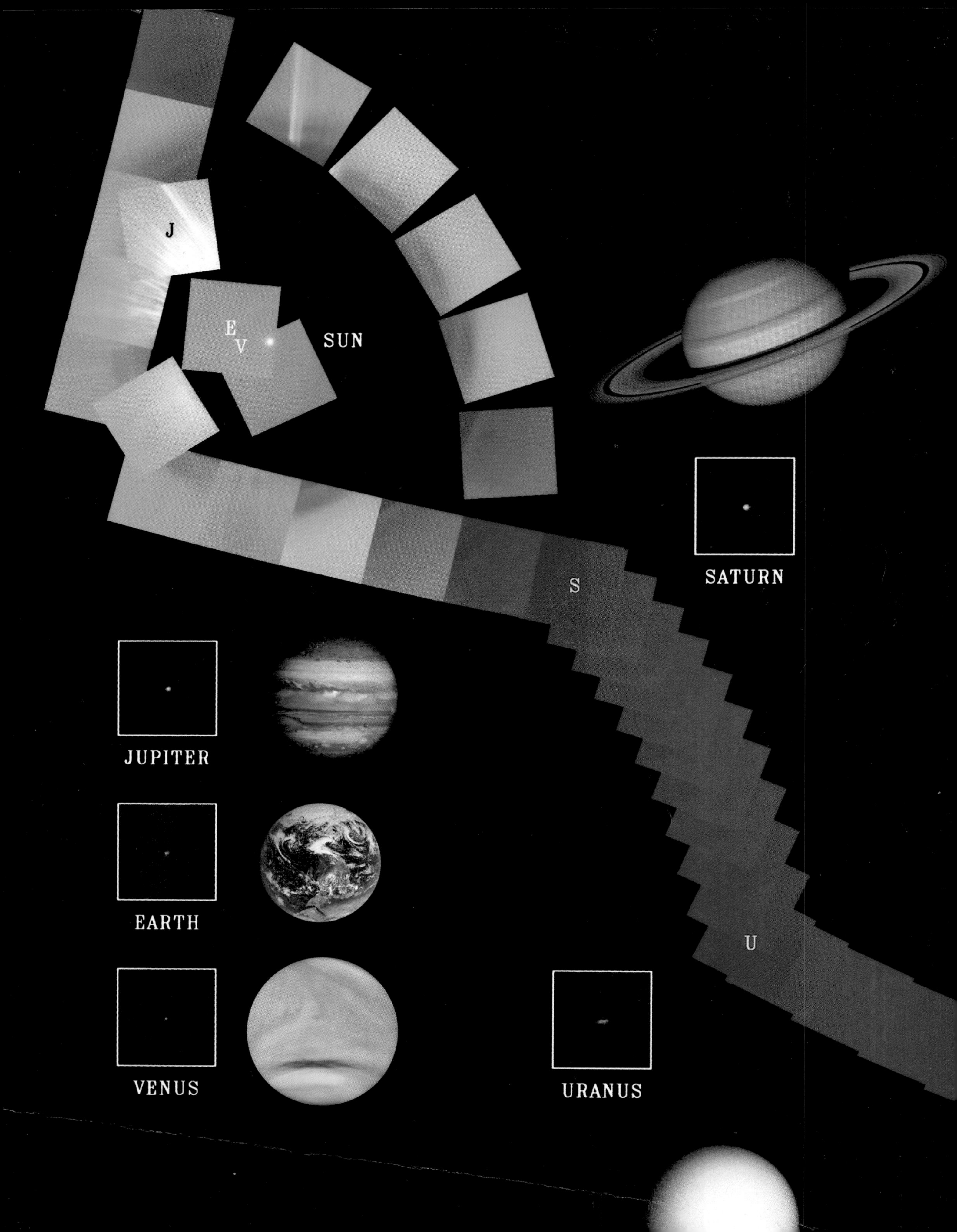
J
E
V
SUN
SATURN
S
JUPITER
EARTH
U
VENUS
URANUS

FAMILY PORTRAIT

Voyager 1 did not join its sister on its trip to Uranus and Neptune. The two spacecraft parted ways at Saturn. *Voyager* 1's last regular assignment was to observe the moon Titan up close. In doing so, it was pulled so powerfully by Saturn's gravity that its trajectory swung upward. It then began to send back new data from the unexplored region high above the planets. It even sent back a unique view of the Solar System itself. On February 14, 1990, *Voyager* 1 took a group photograph from its lofty position nearly six billion kilometers (four billion miles) above the planets. Mercury and Mars were lost in the dazzling Sunlight, and tiny Pluto couldn't be found anywhere. But the other six planets were there. As befitted the end of such an extraordinary mission, *Voyager* 1's parting gift to the people of Earth was a "family portrait" of the planetary system that is their home in a universe of infinite vastness.

Then its cameras were turned off forever.

NEPTUNE

N

5

The Planet That Travels Sideways

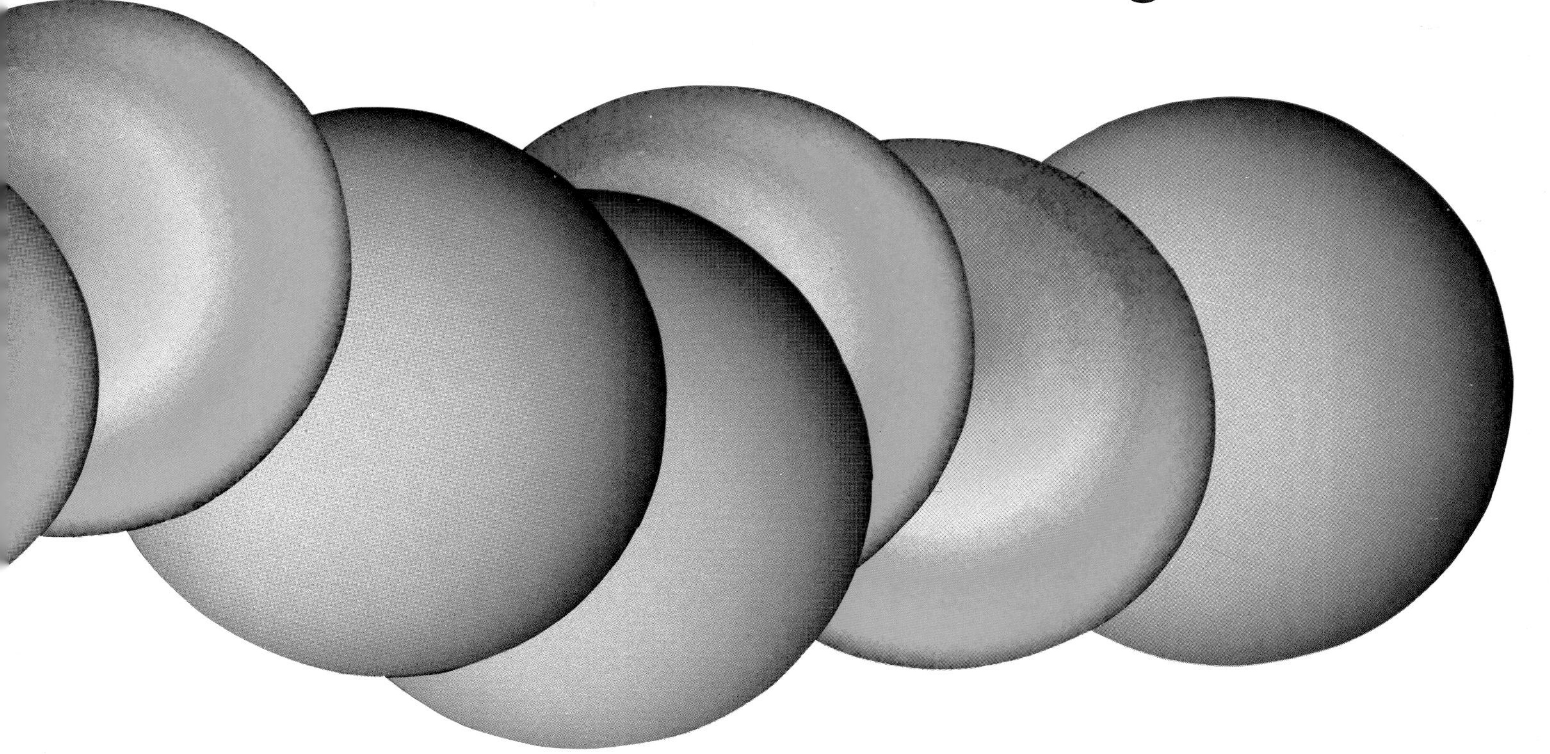

Uranus is so far away from Earth that it takes radio signals and light, traveling at 300,000 kilometers (186,000 miles) a second, two hours and forty-five minutes to get from there to here. It took *Voyager* 2 eight and one half years to get from here to there. If you were standing on Uranus, the Sun would look like a large dot.

Seen through *Voyager* 2's camera on November 4, 1985, the planet looked like a fuzzy blue tennis ball. Unlike Jupiter and Saturn, Uranus did not seem to give off radio signals, and there was no evidence of a magnetic field. "If it weren't for the pictures," one scientist complained before closest approach, "we wouldn't even know that planet is there."

Since Uranus receives only about 1/400 the amount of sunlight that reaches Earth, pictures needed long exposure times. The cameras' shutters had to stay open for 15 seconds or more when photographing the moons. The longest exposure time, 96 seconds, was used to get a dramatic backlit view of Uranus' rings. In order to prevent blurring during such long time exposures, *Voyager* 2's controllers used its thrusters to offset the wobble of the moving spacecraft.

But taking the pictures was just the beginning. Because Uranus is so far, the signal was very weak. Data sent back to Earth had to be slowed down in order to keep it clear. You may do the same thing when you speak more slowly to someone on the other side of a large room when there is a great deal of background noise.

Because the data had to be sent slowly, *Voyager* 2's computer engineers came up with a technique for compressing it. Also, the receiving antennas on Earth were enlarged to collect more radio waves, and their shape was changed to make them more bowl-like. Changing the antennas' shape was like cupping your hand around your ear to collect more sound.

In the four and a half years it took *Voyager 2* to travel from Saturn to Uranus, the mission team made it a better spacecraft than it had been the day it left Earth.

Uranus's Ring System

Not a Lot

Uranus cannot be seen with the naked eye except on clear, moonless nights. It went unnoticed until William Herschel, an amateur astronomer, discovered it with a homemade telescope. He spotted it on the night of March 13, 1781. At first he thought it was a comet. Herschel—later Sir William—turned from amateur to one of the giants of astronomy overnight.

During the next 200 years, astronomers studied Uranus intently, but the planet revealed very little. It appeared turquoise, had an atmosphere rich in methane and hydrogen gases, and was surrounded by five medium to small moons and what appeared to be at least nine black rings. The rings were not seen directly, but astronomers knew they were there because they were in the way of the stars behind them, making the stars twinkle.

No one could tell whether Uranus had a magnetic field. And strangest of all, unlike any of the other known planets, Uranus was tipped on its side. It seemed to rotate vertically—spinning from top to bottom—rather than on the plane of its revolution around the Sun. Its moons circled in the same way.

And to make matters even more bizarre, Uranus rotated backward.

Uranus remained an intriguing mystery until late January 1986, when *Voyager* 2 began its near encounter.

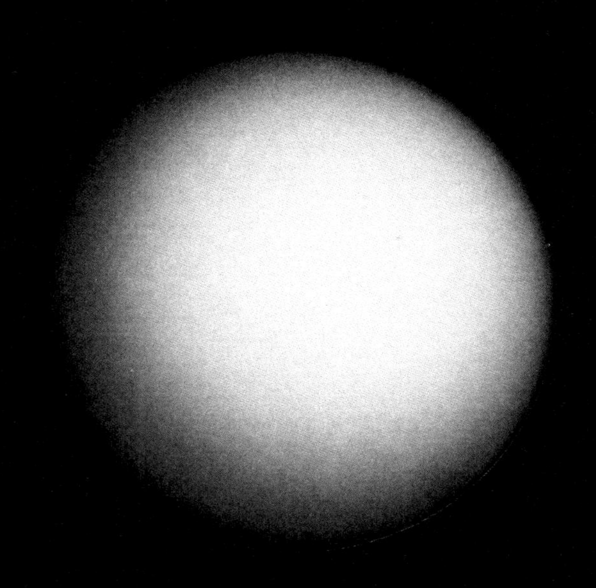

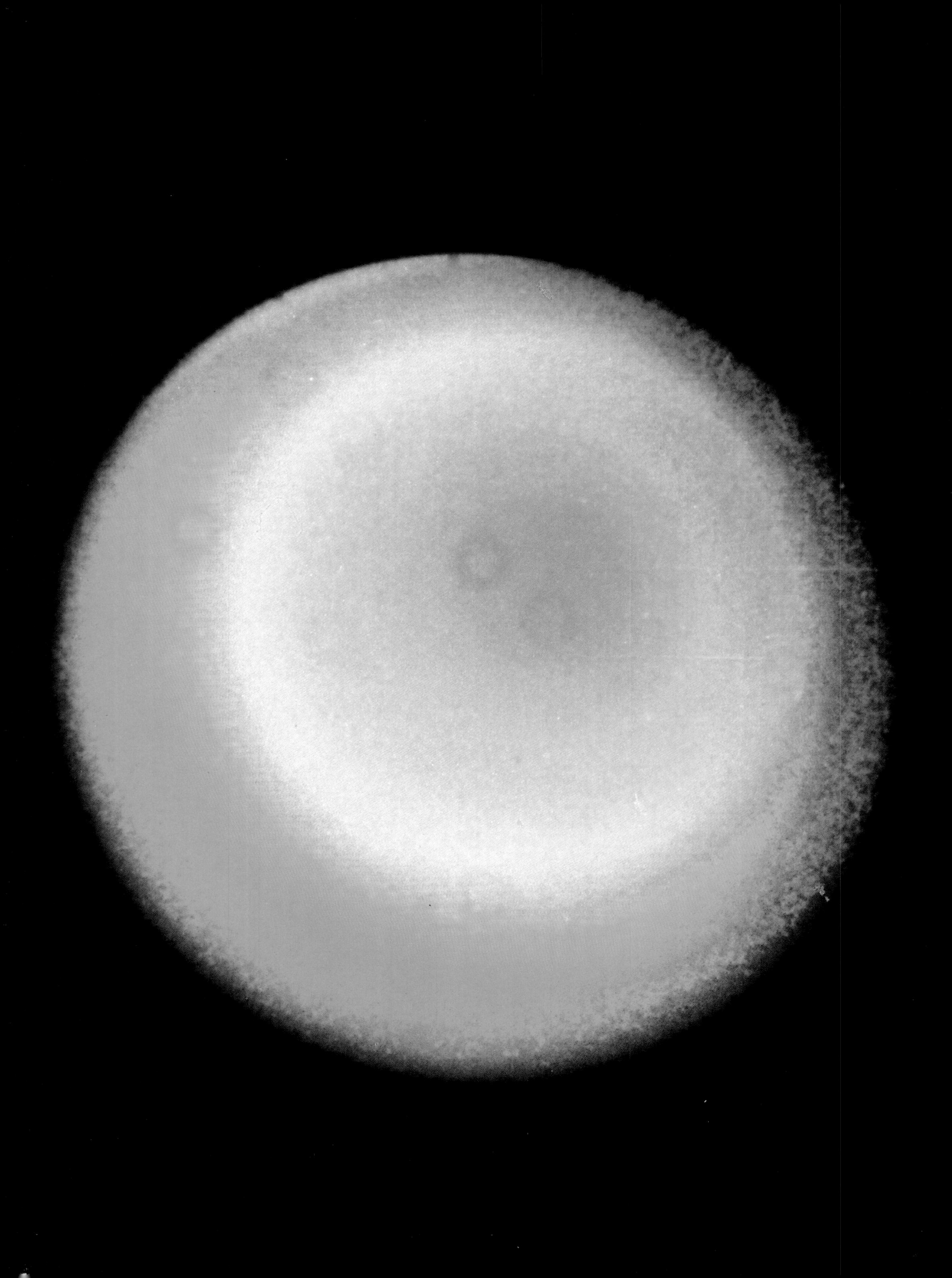

What We LEARNED

Voyager 2's first major discovery was a magnetic field surrounding Uranus, as strong as those around Earth and Saturn, though weaker than Jupiter's. We think magnetic fields are caused by fluids that conduct electricity—for example, the molten iron inside Earth. *Voyager* data led mission scientists to theorize that the Uranian magnetic field is caused by an electrically conducting, superheated layer of water above a rocky, molten core.

Voyager 2 discovered Uranus's atmosphere to be composed mostly of hydrogen and helium. About two percent of the atmosphere right below Uranus's clouds is methane, a major component of natural gas and the same chemical that is sometimes used on Earth to propel automobiles. Methane gas in the Uranian upper atmosphere absorbs, rather than reflects, red light. That leaves blue and green light to give the planet its distinctive turquoise color.

Uranus's rings are held together by shepherd moons

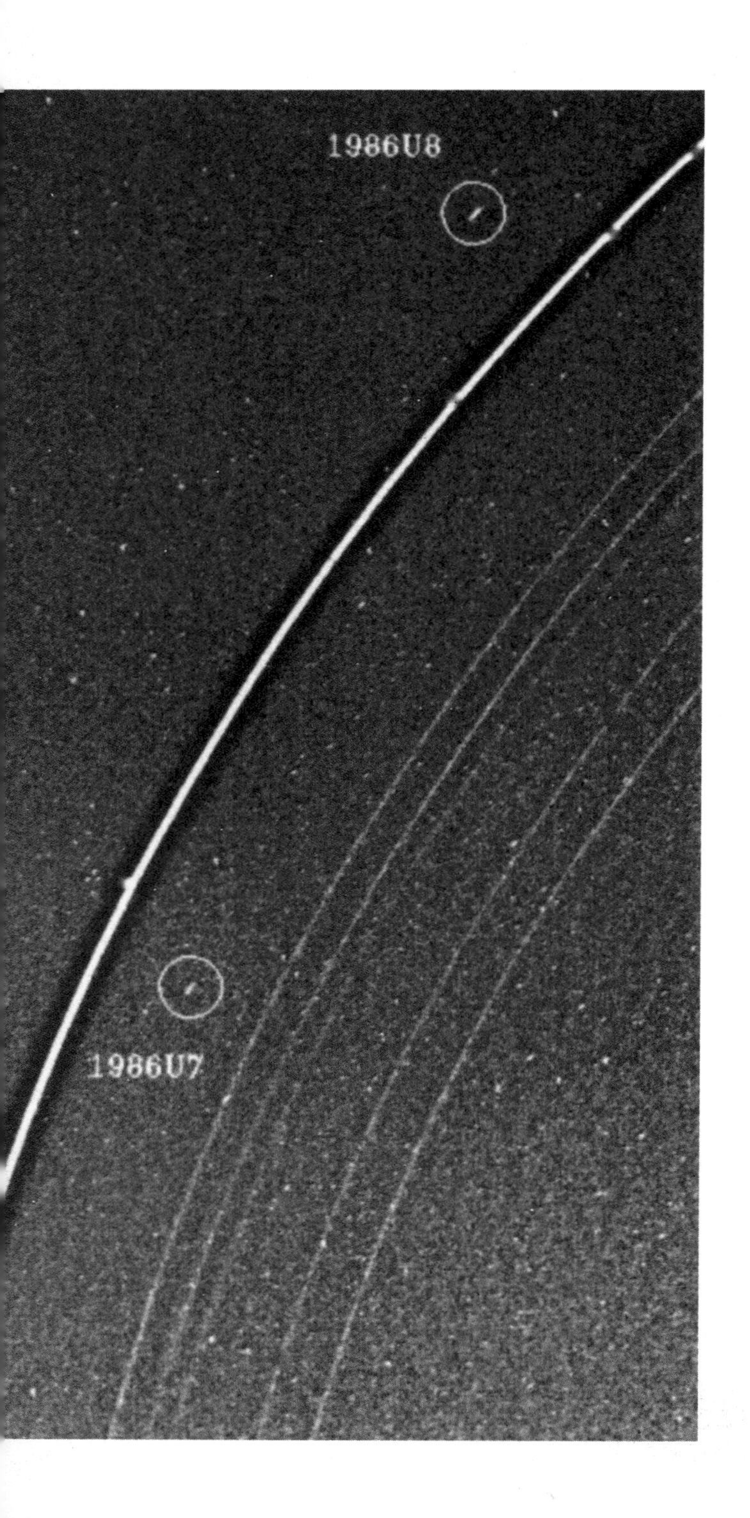

Then there were Uranus's elusive rings. Nine of them, some as black as coal, were spotted on the approach. Now all nine were photographed and measured, and two new ones were found. Unlike the rings of Jupiter and Saturn, which are mainly made of dust and sand-size material, the Uranian rings include some quite large objects. Much of the structure of Epsilon, the outermost ring, was seen to consist of boulders, or perhaps boulder-size chunks of ice, several feet across. Shepherd moons seem to keep the rings together through effects of gravity.

As the near encounter ended, scientists could still only guess at why Uranus travels sideways in its 84-year orbit around the Sun. They think that something nearly as big as the planet itself struck it a terrific glancing blow at some point and then continued on its way. But no information provided by *Voyager* 2 could prove or disprove that theory.

As the spacecraft headed toward Neptune, the sun lit the back of Uranus, allowing *Voyager 2* to take one of its most famous images—the Blue Crescent.

SATELLITES

The speeding spacecraft sent back to Earth information about ten previously unknown satellites of Uranus, along with data about the five already known satellites. Because of its enormous speed and the steep tilt of the Uranian system, mission scientists had less than a day—January 24, 1986—to collect in essence all the detailed information *Voyager* 2 would ever give them on the moons. What they had thought were large chunks of ice when seen as specks of light through telescopes became five distinct worlds under close-up inspection.

Although all five major satellites seemed to be made of about 50 percent water ice, 30 percent rock, and 20 percent carbon- and nitrogen-based materials, they appeared to be very different from one another in other respects.

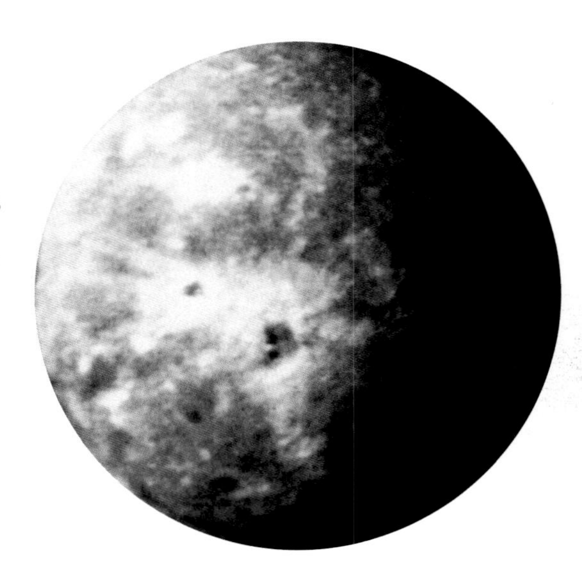

Miranda has three oval-shaped regions that contain a crazy quilt of ridges, terraces, and fault canyons as deep as 12 miles.

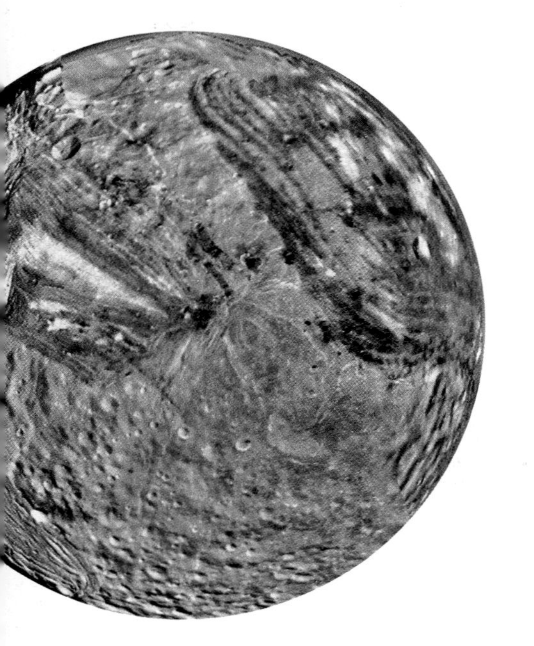

Oberon has a tall mountain and craters filled with hardened lava on its icy surface.

Ariel has the brightest and perhaps the youngest surface in the Uranian system and relatively few craters. Apparently some kind of low-velocity material has hit the surface and, like sandpaper, rubbed away the larger craters. Ariel's many faults and flows of icy material indicate that it has a turbulent history.

Umbriel is old, dark, and heavily cratered. It shows little evidence of geologic activity.

Titania has large fault lines and canyons. These are evidence of earthquakes, which indicate activity beneath the surface.

6
At Neptune. . .

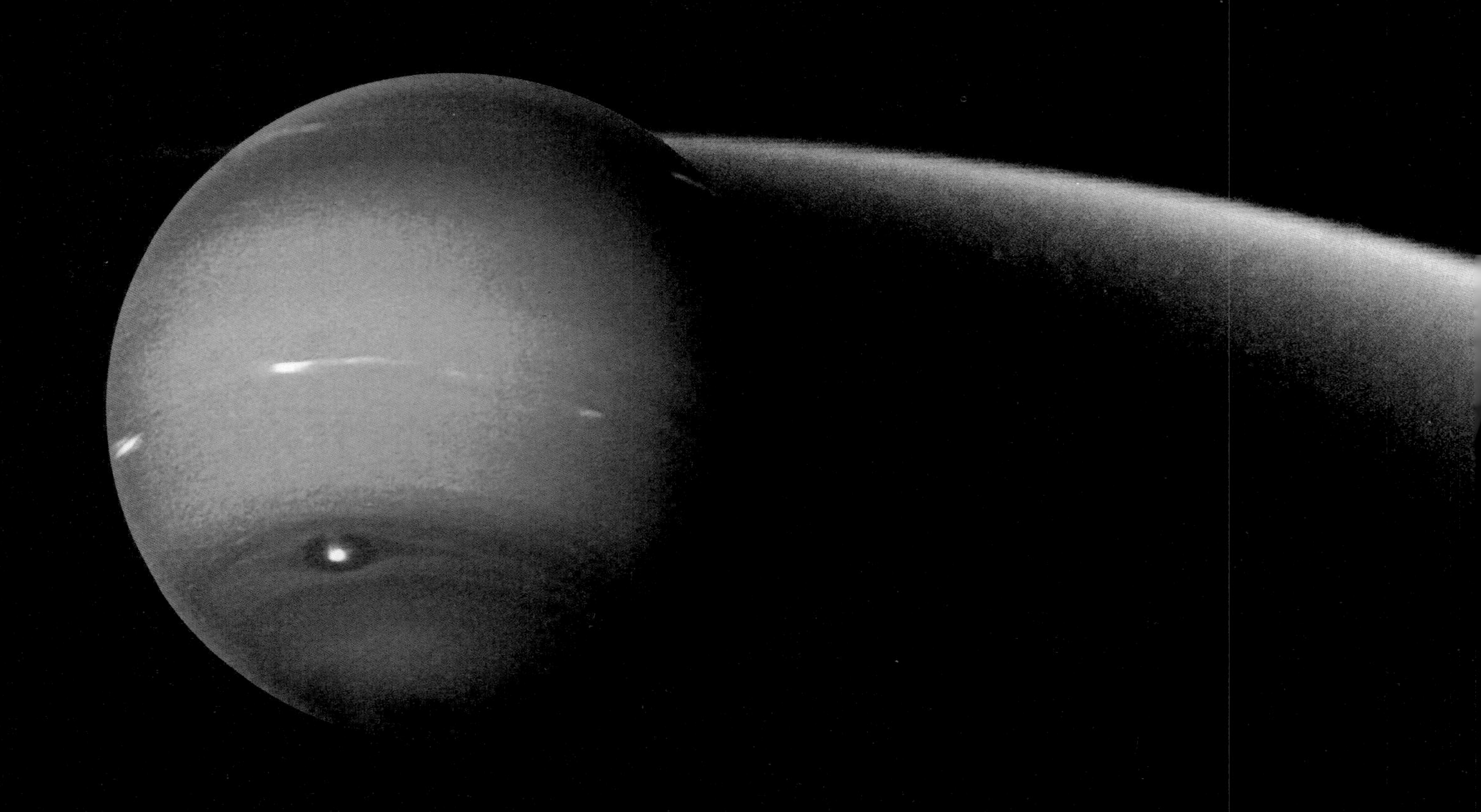

The Last Pictures

Voyager 2 made its last planetary port of call on the night o
August 24, 1989, when it skimmed above Neptune's blue
methane-shrouded atmosphere. The distance from Earth tc
Neptune is so great that it took four hours and six minutes for the
data *Voyager* 2 collected to reach JPL.

Having flown some 6.5 billion kilometers (4 billior
miles) in 12 years, *Voyager* 2 arrived at Neptune within 3
kilometers (20 miles) of where it was aimed, and 2.
seconds of its scheduled arrival time. It was ar
unequaled triumph for the navigators.

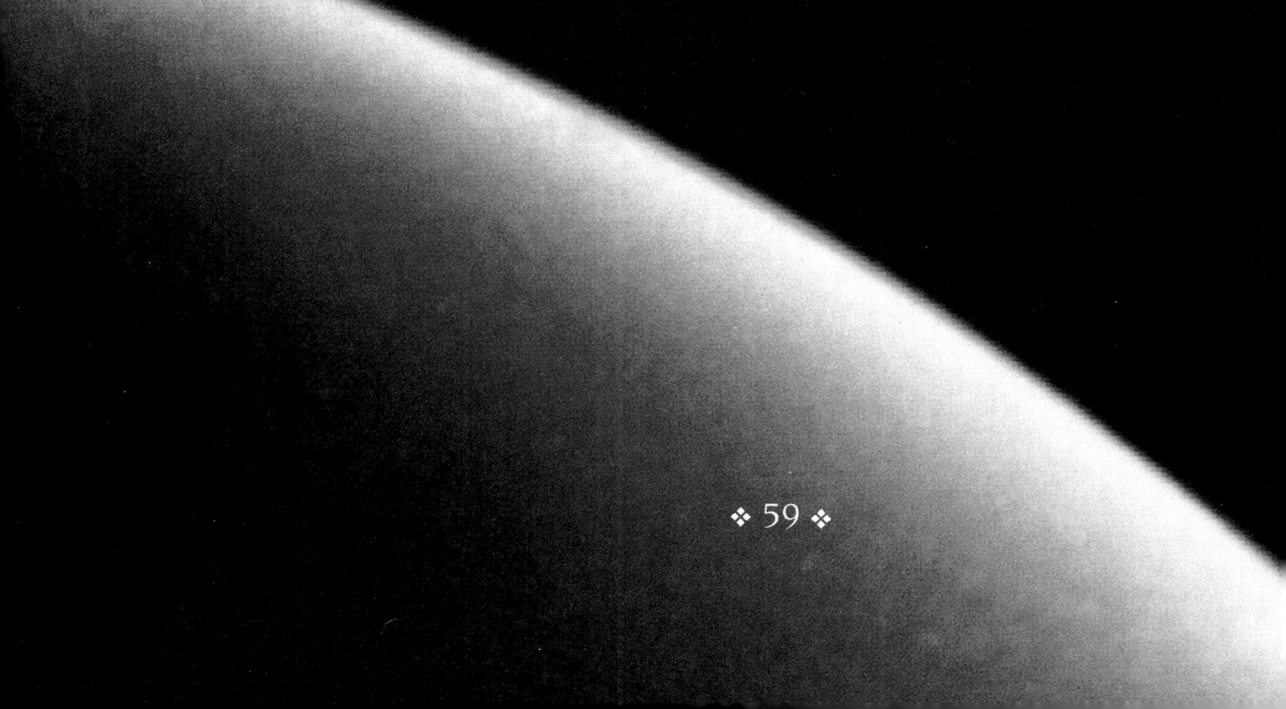

Neptune is so far away that it is the only planet to be dis-
vered by mathematical calculation instead of direct observa-
on: it was deduced before it was seen. And its presence was
ven away by Uranus.

By the mid-1840s astronomy had become so precise that
anets, stars, and other celestial bodies could be tracked close-
An English mathematician named John Couch Adams and a
ench mathematician named Urbain Jean Joseph Leverrier,
orking independently, noticed that Uranus's orbit was most
eculiar. Records showed that Uranus had sped up slightly

before 1822 and then slowed down again. A large object seemed to be tugging at Uranus and affecting its motion. A close look at the part of the sky where the object should have been revealed Neptune—the eighth planet.

Nearly 4.5 billion kilometers (2.8 billion miles) from the center of the Solar System, Neptune whirls in about one thousandth of the light that falls on Earth. It is invisible to the naked eye from Earth and, since it was discovered, has yet to complete its 165-year orbit around the Sun.

Before *Voyager* 2's arrival at Neptune in August 1989, the only view we had of the planet, even through the best telescopes, was that of a fuzzy speck. It seemed to be a bit smaller than Uranus, with light regions toward the poles—probably ice crystals—and cloud formations around its middle. In fact, Neptune resembled Uranus. Like Uranus, it was thought to have a heart of rock and ice covered by fluid. Unlike Uranus, Neptune had only ring arcs—partial, incomplete rings. If that's what they were, they would definitely prove to be some of the weirdest objects in a solar system already loaded with weird objects.

Finally, two moons had been seen orbiting Neptune. The larger, named Triton, was thought to be about the size of Earth's Moon. Triton was already known to move in a peculiar, retrograde orbit—in the direction opposite Neptune's spin—but no one knew why. The other, much smaller, satellite was named Nereid.

What We LEARNED

In contrast to Uranus's bland face, Neptune proved to be a wild, wonderful world. Its winds were even more ferocious than those in Jupiter's renowned cyclones.

There was an Earth-size hurricane named the Great Dark Spot (after Jupiter's Great Red Spot). A white super-squall named the Scooter whipped around the planet every 17 hours. *Voyager* 2's cameras spotted parallel streaks of high-level cirrus clouds that cast their shadows on the thick blue cloud deck some 50 kilometers (30 miles) below. Such clouds, rare in the Solar System, indicate that Neptune's atmosphere is even more dynamic than Jupiter's.

Voyager 2 showed pictures of four complete rings and six new moons. From Earth, the rings had appeared to be arcs because some sections were thicker than others. The new moons—chunks of dark rock—were named Naiad, Thalassa, Despia, Galatea, Larissa, and Proteus. The first good pictures of Nereid showed that it looked like a dirty potato the size of California.

Voyager 2's encounter with Neptune was unique in two respects. First, it gave the inhabitants of planet Earth the best look they had ever had at a distant world. And second, it marked the official end of an extraordinary odyssey. Starting when the mission was conceived and the experiment planned, many of the scientists, engineers, administrators, and support personnel at JPL and the various participating universities had spent 17 years on the project. They were always mindful of the fact that they were participating in one of the great feats of adventure and exploration of all time. They therefore learned to work smoothly together as the immense flow of data from both *Voyager*s streamed back home in an uninterrupted **whisper over** enormous distances.

Voyager 2 sailed silently over Neptune on the night of August 24. Then it swung down behind the blue planet and streaked past Triton at a distance of only 38,000 kilometers (24,000 miles). Scientists had been teased during the weeks before near encounter by steadily growing pictures of the pinkish ball with its complicated surface. The fact that Triton circles Neptune opposite to Neptune's spin indicates that it once traveled on its own but somehow came too close to Neptune. Perhaps it collided with one of Neptune's moons and was snared by the large planet's gravity. It then became a mere satellite. But what a satellite!

Triton is a little smaller than Earth's Moon. With polar caps crusted with methane and nitrogen, it is the coldest place ever measured in the Solar System. Still, Triton is tilted 20 degrees off its rotational axis, so it has seasons the way Earth has seasons. Triton's south pole was at the height of its 41-year-long summer when *Voyager* 2 passed by, except Triton's summer certainly isn't what we'd call beach weather.

Part of Triton's surface is a tortured landscape of huge canyons, frozen lakes, sharp breaks and tears, craters, and mysterious dark streaks. The streaks make its surface look as though it had been shot by some gigantic gun that caused some of its inside to leak out. The streaks were eventually seen to be caused by plumes coming from beneath Triton's surface. Scientists first theorized that the plumes were gas and dust being vented by the moon. In 1992, scientists poring over the same pictures noticed volcanoes—not lava-spewing ones like those on Earth but volcanoes that gushed ice.

Another part of Triton looks like the webbed skin of an enormous cantaloupe. Scientists concluded that this smoother, young part of Triton was formed by moving streams of ice. The webbing was caused by sheets of ice or rock coming together and crunching into each other. The area where Triton's two regions meet is bluish, making it the only moon in the Solar System known to have that color.

The two *Voyagers* sent back so much data about the four giant outer planets, their moons and ring systems, and the interplanetary space between them that scientists will be kept busy analyzing it all, well into the twenty-first century.

7
Voyage to Forever

These are among the highlights of the Grand Tour:

Jupiter's swirling Great Red Spot was examined in detail, and lightning was seen in the planet's atmosphere. Winds recorded on Saturn were found to be the fastest in the Solar System.

The *Voyager* mission team discovered a ring around Jupiter, two new rings around Uranus, complete rings rather than arcs around Neptune, and details about the composition and structure of all four planetary ring systems. Saturn's beautiful rings were seen to have "spokes" and be composed of thousands of ringlets, resembling grooves in a phonograph record.

They discovered three new moons of Jupiter, ten new moons of Uranus, and six new moons of Neptune. In addition, many of the known moons were inspected up close for the first time, showing among many other things that Jupiter's Io had active volcanoes. Neptune's Triton is probably active too, with geysers and ice lava.

All the millions of little facts that the *Voyagers* brought back give the people of Earth a clearer picture of what the Solar System is made of and how it works. Perhaps more important, much of the information raises new questions that cannot yet be answered. For the mission scientists, the questions are much more fun than the answers. Working out the puzzle is the important thing in science.

The *Voyager* twins' active journey is still far from over. They are still exploring and sending back data as part of a new assignment: the *Voyager* interstellar mission. Both are sailing gracefully on the solar wind as they search for the outer edge of the Solar System, where the Sun's influence ends and interstellar space truly begins.

Although their video cameras are turned off, other equipment is expected to keep making observations at least until 2020, possibly 2030.

One day, however, both spacecraft will die, as their power sources run out. Unable to send data to Earth or receive steering instructions, each will slowly tumble on a course that will take it to the farthest reaches of this galaxy and perhaps to others beyond. *Voyager 1*'s course is presently set for the constellation Hercules. *Voyager 2* is pointed at a place just to the southwest of the constellation Sagittarius, which, as seen from Earth, is near the center of our own galaxy, the Milky Way.

But the end of the *Voyager* mission will by no means be the end of planetary exploration. Indeed, looked at from the long perspective of history, it is only the beginning.

Glossary

Asteroid belt
An area between Mars and Jupiter filled with thousands of small planetlike objects.

Atmosphere
The blanket of gas that covers a planet or moon.

Boom
A long arm that holds instruments away from the spacecraft's bus.

Booster
The rocket that propels a spacecraft into space.

Bus
The main body of the spacecraft, containing computers, recorders, transmitters, and receivers.

Celestial
Relating to the sky.

Cirrus cloud
A wispy white cloud usually made of tiny ice crystals.

Colossus
A gigantic object.

Constellation
A configuration of stars.

Comet
A celestial body with a bright head and a long tail.

Cosmic ray
High-energy particles that come from outer space and hit other particles in the Earth's atmosphere.

***Deep Space Network* (DSN)**
The worldwide electronic system that uses antennas to communicate with spacecraft.

Downlink
The communication route that is used to send data from a spacecraft to Earth.

Galaxy
One of billions of star systems that make up the universe.

Galilean moons
The moons of Jupiter that were discovered by Galileo—Callisto, Ganymede, Europa, and Io.

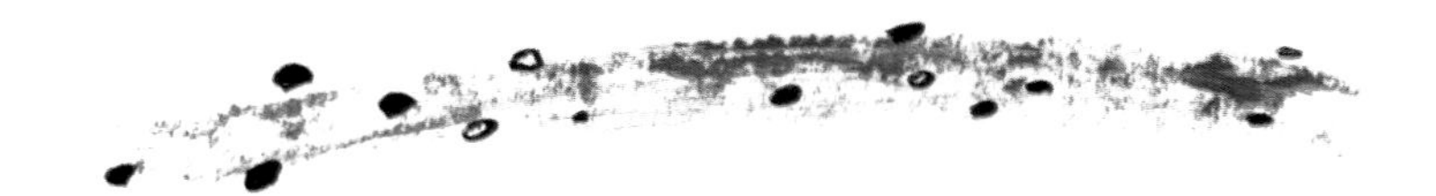

Gaseous
Made up of gas.

Gravity
The attraction of one planet or other celestial body for another.

Gravity assist
The use of a planet's gravitational pull and motion to increase the velocity of a spacecraft. Also called gravity propulsion.

Hypergolic fuel
A propellant that ignites when two fuels make contact.

Impact craters
Holes in the surface of a planet or moon made by something smashing into it.

Infrared
The part of the electromagnetic spectrum between visible light and radio waves. Infrared waves are invisible to the human eye but can sometimes be felt as heat.

Interstellar
Between stars.

Interplanetary
Between planets.

Jet Propulsion Laboratory (JPL)
The science center in Pasadena, California, where the *Voyagers* were designed, built, and controlled.

Jettison
Throw away.

Jovian
Relating to Jupiter.

Lunar
Relating to the Moon.

Magnetic field
The region of space surrounding a planet in which objects can be subjected to a magnetic force exerted by the planet. The force is the strongest near the planet's magnetic poles.

Mass
The amount of material an object contains. Mass can be measured as weight when there is gravity pulling on it, as on Earth.

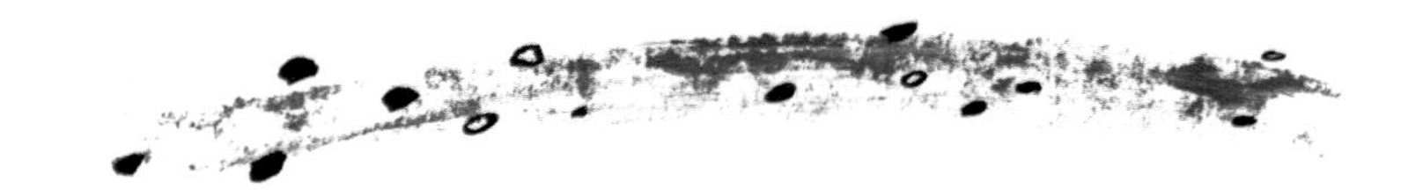

Milky Way
The galaxy of which the Sun and Solar System are a part.

Moonlets
Small moons.

National Aeronautics and Space Administration (NASA)
The United States civilian space agency.

Orbit
The path of one celestial body around another.

Photo polarimeter
An instrument for determining the direction light waves are oscillating in.

Planetary system
A group of planets and their moons that circle the same star.

Plasma
Charged particles that are so thick they form a gas.

Plutonium
A radioactive metal that gives off energy as it decays.

Propellant
A fuel used by rocket engines.

Propulsion
The force that makes machines such as jet planes and spacecraft move.

Rendezvous
Come together.

Resolution
Clarity of focus in a photographic image.

Ring arc
An incomplete ring shaped like a horseshoe.

Ring system
A planet's rings.

Rocket
An engine that can operate in space because it carries its own fuel and oxidizer.

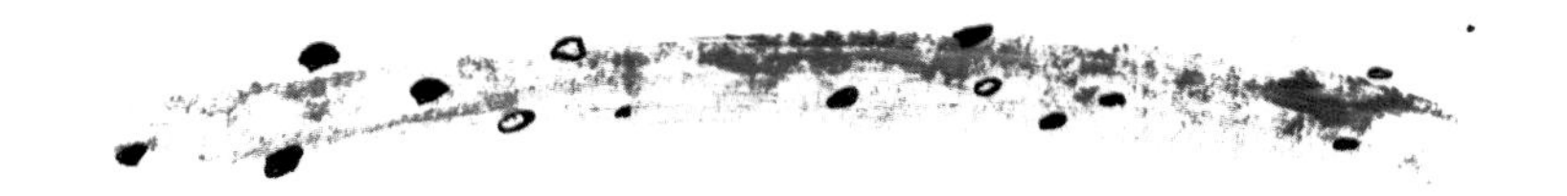

Satellite
Any object that orbits another object. Earth is a satellite of the Sun; the Moon is a satellite of Earth.

Shroud
A fiberglass guard that protects the spacecraft from heat.

Solar cell
A unit that converts sunlight into energy. Panels full of such cells power spacecraft in Earth orbit and as far away as Mars.

Solar wind
Particles radiating from the Sun in all directions and traveling at about one million miles an hour.

Spectroscope
An instrument that analyzes the wavelengths radiated by objects to determine what they are made of and how hot they are.

Superheated
Heated above the usual boiling temperature but prevented from actually boiling.

Thruster
A small rocket-type engine that helps steer a spaceship.

Ultraviolet
Electromagnetic radiation with wavelengths shorter than those of violet light but longer than those of X rays. It is not visible to the human eye but causes sunburn.

Unprecedented
Never before seen.

Uplink
The communication route that is used to send instructions from Earth to a spacecraft.

Velocity
The speed of an object in a given direction.

Index

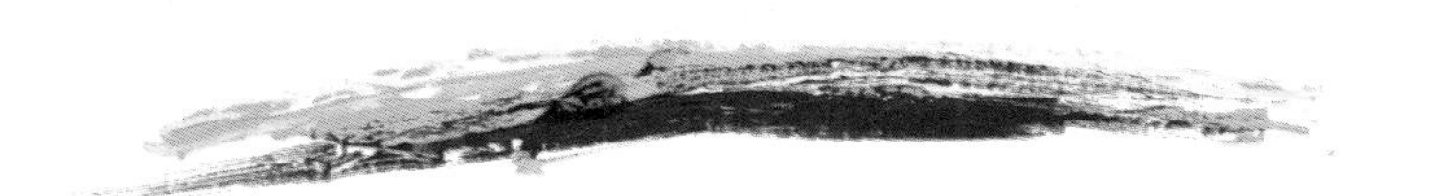

For Further Reading

* *means especially suitable for young readers*

Burrows, William E. **Exploring Space: Voyages in the Solar System and Beyond.**
New York: Random House, 1990.
This book talks about the politics and science that made space exploration possible and about the remarkable accomplishments of the U.S. space program.

*Chartrand, Mark R. **Planets: A Guide to the Solar System.**
New York: Golden Press, 1990.
A handy pocket reference book filled with facts about the Solar System.

*Curtis, Anthony R., ed. **Space Almanac, 2nd edition.**
Houston, TX: Gulf Publishing Company, 1992.
Everything you could ever imagine wanting to know about space, and more.

*Gibson, Bob. **The Astronomer's Sourcebook.**
Rockville, MD: Woodbine House, 1992.
A listing of planetariums, space museums, space camps, and space clubs all over the United States.

Henbest, Nigel. **The Planets: A Guided Tour to the Solar System Through the Eyes of America's Space Probes.**
New York: Viking, 1992.
If you liked Mission to Deep Space *and want lots more detail, try this one.*

*VanCleave, Janice. **Astronomy for Every Kid: 101 Easy Experiments that Really Work.**
New York: John Wiley & Sons, 1991.
This book is full of experiments that are fun to do and help you understand how we know what we know about space.

Picture Credits

All of the photographs in this book are courtesy of NASA/JPL (Jet Propulsion Laboratory) with the following exceptions: Bettman Archives, 6, 7; Lowell Observatory, 36-37, 50; Mauna Kea Observatory, 26-27; Jim Riffle, 72-73; Royal Observatory, Edinburgh/ Anglo-Australian Telescope board, 10-11. Artwork by Robert Caputo, 12-13; Shelley Pritchett, 31; Steven Sullivan, 42-43; Gary Tong, 18-19, 20, 21.